Trespassing

Other books by John Hanson Mitchell

Ceremonial Time

A Field Guide to Your Own Back Yard

Living at the End of Time

Walking Towards Walden

Trespassing

*An Inquiry into the
Private Ownership of Land*

John Hanson Mitchell

A Merloyd Lawrence Book

PERSEUS BOOKS
Reading, Massachusetts

Library of Congress Catalog Card Number: 99-61915

Perseus Books is a member of the Perseus Books Group.

Find us on the World Wide Web at
http://www.aw.com/gb/

Cover design by Suzanne Heiser
Text design by Greta D. Sibley & Associates
Set in 11.5-point Cheneau by Greta D. Sibley & Associates

1   2   3   4   5   6   7   8   9—MA—02   01   00   99

For James and Hugh,
who taught me the art

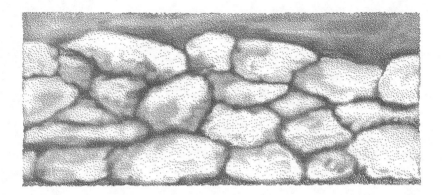

Ona woh nag pish nootamwog kummithe wetanonk, kah kummrnueu Israel....
(I have been a stranger in a strange land.)

—*Exodus 2:22,* Up Biblum God, *translation of the Bible
into Algonquian by John Eliot, 1661*

Contents

Acknowledgments..xiii

Chronology..xv

Players..xix

Prologue..1

ONE: A Certain Tract of Land...7

TWO: Owners and Outcasts..19

THREE: Should Trees Have Standing?...31

FOUR: The Cords of Christ's Tent..49

FIVE: To Have and to Hold..61

SIX: Common Ground...75

SEVEN: Cross-Lot Walking...89

EIGHT: Terra Nullius..105

NINE: Holding Ground..121

TEN: Out of the Quiver of the Scriptures...................................133

ELEVEN: The Last of the Commons...149

Contents

TWELVE: Islands of the Dead 165

THIRTEEN: Who Really Owns North America? 181

FOURTEEN: The Tawny Vermin 193

FIFTEEN: Their Heirs and Assigns Forever 213

SIXTEEN: The Intelligence of Salamanders 237

SEVENTEEN: The Landing 253

EIGHTEEN: Drawn and Quartered 269

Epilogue 285

About the Author 295

Acknowledgments

Quite apart from a lot of snooping around places where, by law, I had no right to be, this book involved prolonged discussions with any number of authorities in a variety of fields. I would like to acknowledge them, as well as the individuals portrayed in the narrative, for their time, information, entertainments, and support. I want to thank Shirley Blanke, Colin Calloway, Linda Coombs, and John Forti for their help with seventeenth-century Native American customs; Sam Sapiel of the Penobscots, and Gary McCann and members of the Muhucanuh Confederation, for their Deer Island work; Michael Roberts and Georgess McHargue, for their help with the archeology of the Nashobah site; Mark Strohmeyer, James Mavor, and the artist Marty Cain, for information (or opinions) on the spiritual overtones of the landscape around Nashobah; Morse Payne, for his detailed information on seventeenth-century mapping; Sandy Williams and Barbara Nylander, for their help with the story of the Jones family and the last days of Sarah Doublet; Ivan Masser, for La Herencia story; Don and Carrie Prouty, for the stories of the old house at Nagog Pond (not to mention the use of their landing at Long Lake); Doreen and Bob Morse, and Marie Lombardo, for the Nagog Hill property information; and Linda Clark and the crew at Nagog Hill Orchards for putting up with my eternal questions about apple growing. I also want to thank Eunice Morrison, owner of the orchard; I hope she will forgive my trespasses. And I want to acknowledge the help of David Koonce, and especially the work of Sara Foss and Patty Townley.

They put in as much time, energy, and tears into the salvation of the Frost Whitcomb tract as the hero of this account, Linda Cantillon.

For background on American legal matters I want to thank Steve Ells, Henry Brown, and Dick Emmett; Andrea Simoncini of Florence, for his information on Roman law and the Cadore district in northern Italy; and Tamar Karet for tips on current British land use issues. I also want to acknowledge my cousin James Andrew Mitchell for his fine written account of the doings at Thainstone in Kintore; Julian Crandall-Hollick of Independent Broadcasting Associates for allowing me to use some of the material from the transcripts of the National Public Radio program "Trespass" (as well as the many bottles of good Burgundy); Lawrence Millman for his monster stories and his feral mushrooms; and for support, inspiration, patience, and, one might even say, their endurance, Jill Brown, Carol Towson, Robie Hubley, Merloyd Lawrence, Steve Hecht and Dori Smith, Lansing Old, Ellie Bemis, her father John Bemis (whose diatribes on the nature of title were an inspiration and an indication that I was on the right track), and everyone else who in any way helped convince me that a book involving law would not be profoundly boring.

Finally, I would like to thank those courageous few, past and present, who have possessed the wisdom, benevolence, foresight, and generosity to donate or otherwise transfer their private lands to public use.

Chronology

For the most part this story takes place in two different time periods: the late seventeenth century and the late twentieth century. Some thirty generations of Indians, English, and Americans have lived on the site in the time of this story, and even though I've conveniently skipped some two hundred years of occupancy (between 1736 and 1996, more or less), the times and the places and the people can get confusing. Thus this chronology:

1215 Magna Carta. Right of the king to appropriate property from the nobles or the church limited.

1381 Wat Tyler's Rebellion. Peasants attempt to free themselves from the feudal system. Suppressed.

1500 The *encomienda* is established in the Americas. Natives are owned as property.

1633 Beginning of the "Great Migration." Some three thousand Puritans settle in New England over the next two to three years.

1637 English buy the six-square-mile tract of land later named Concord from the local Pawtucket Indian leaders Tahattawan and Squaw Sachem.

1646 John Eliot, the so-called Apostle to the Indians, begins missionary work among the New England Indians.

1651 Eliot converts Waban, establishes the first Christian Indian village at Natick.

1654 Eliot founds seven more Christian Indian villages, including Nashobah Plantation.

1660 Mohawk raids on Nashobah Plantation; village is deserted for a year.

1666 Building codes established in London as a result of the Great Fire.

1668 Earthquake at Nashobah village site.

1675 King Philip's War begins.

1675 (August) General Court issues proclamation confining all Indians to the Christian Indian villages.

1675 (Autumn) Deer Island internment camp established.

1675 Nashobah Indians rounded up and marched to Concord.

1676 (February) Nashobah Indians transported to Deer Island.

1676 (Spring) King Philip's War ends. Surviving Indians are released from Deer Island.

1676-80 Remnant Nashobah people living at the plantation.

1686 First survey of Nashobah Plantation.

1687 Two-mile swath cut out of the plantation and sold to two Groton families, Lawrence and Robbins.

1694 English purchases cut the plantation by half.

1732 Jones Tavern constructed.

1734 Ephraim and Elnathan Jones take over care of Sarah Doublet, the last surviving Nashobah Indian.

1736 (September) Sarah Doublet signs deed to the Jones family for the last five hundred acres of Nashobah Plantation.

1736 Sarah Doublet dies; Joneses assume ownership.

1756-1996 Nashobah Plantation site sold, resold, cut over for timber, farmed, sold again for dairy farms, sold again to summer cottage people, reforested, sold again in one-acre lots for permanent houses, cut again, restored to orchards, reforested a second (or third) time.

1785 The Land Ordinance. Western territories surveyed into a grid.

1791 Fifth Amendment to the U.S. Constitution.

1796 The Calico Uprising. Farmers on the vast Van Rensselaer estate along the Hudson River organized resistance to gain their own land.

1805 *Pierson* v. *Post.* Ownership of wild nature established by courts.

1887 Dawes Act. Common lands of the Native American people privatized.

1926 *Village of Euclid* v. *Ambler Realty Co.* U.S. Supreme Court allows the use of zoning codes.

1964 John Morrison begins buying tracts west of Nagog Pond, sells dairy herd, and begins planting apples.

1992 Littleton Conservation Trust assumes control of ninety acres of original village site on the hill west of Nagog Pond; the common land is restored.

1995 One hundred and thirteen acres west of Fort Pond proposed as housing development.

1995 (Spring) Friends of Open Space contest the sale.

1997 Town of Littleton purchases tract as open space.

Players

In chronological order—more or less.

Antecedents

The Green Man: A benign forest-dwelling figure, half animal half man, who dressed in ivy and protected the straying peasants in medieval Europe. His roots go back to Sumer and Mesopotamia.

Jesus Christ: Questioned the idea of private property. Later revered as a god (one of many) by the Nashobah people.

Justinian: Sixth-century AD emperor who codified Roman Law.

William the Conqueror: Norman warlord, the last invader of England. Declared the English forests his property and evicted the peasants. Ordered the Domesday Book, the accounting of his holdings in England.

William Rufus: His son and successor. Killed in the forest while hunting deer.

Robin Hood: Thirteenth-century, semi-legendary descendant of the Green Man who lived in the forest and questioned the right of *anyone* to appropriate land.

Wat Tyler: Fourteenth-century leader of a peasant revolt that sought to redistribute property.

INDIANS AND ENGLISH

Tahattawan: Sachem of the Pawtucket people. Sold "his" land to the English. Great-grandfather (perhaps) of Sarah Doublet.

John Eliot: Minister at Roxbury, Massachusetts, and the so-called Apostle to the Indians. Converted members of various New England tribes to Christianity and established "praying towns" where his converts could live unmolested by the English. A tireless worker for his cause, unpopular in his time (because he accepted Indians) as well as ours (because he broke their culture).

Daniel Gookin: English soldier and assistant to Eliot. A protector of native people and early chronicler of their culture.

Waban: A sachem of the Nipmuck tribe in central Massachusetts. Eliot's first and most powerful convert. Converted many of his own people.

John Tahattawan: Son of Tahattawan. A lackluster Christian at best. Grandfather (?) of Sarah Doublet.

Sarah Doublet: Also known as Wunneweh, also called Kehonosquaw, the last of the Nashobah people and the last owner of the land at the praying town, Nashobah Plantation. May have lived to be one hundred and may have purposely protected the village site because it was held sacred by her traditional religion.

Tom Doublet: Also known as Nepanet; the last of Sarah's three (?) husbands.

John Speen: A Natick Indian associated with the village at Nashobah. Laid claims to the tract after King Philip's War.

John Thomas Good Man (or Goodman): The Indian teacher or minister to the people at Nashobah at the time of King Philip's War.

Crooked Robin: A Nashobah Indian, possibly a former medicine man. (He may have been epileptic or crippled.)

Samuel Mosely: A soldier of fortune, hired by the General Court to fight Indians during King Philip's War. Evicted the people at Nashobah.

Samuel Hoar: Protected (at his own expense) the Nashobah people at Concord for a few months.

Elnathan Jones; Ephraim Jones: Cousins from Concord, Massachusetts, who took on the care of Sarah Doublet in her last years. They acquired the five-hundred-acre tract after her death.

AMERICANS

Big Thunder: A white man who played the role of an Indian and led the Calico Boys in the Rent Wars in the Hudson River Valley after the American Revolution.

Emiliano Zapata: Early-twentieth-century leader of land revolts in Mexico, sought to return the former common lands (*ejidos*) to the people.

John Adams Kimball: Dairy farmer and landholder on the western edge of the Nashobah tract. Sold land to the Littleton Land Trust. Patriarch of the Kimball clan.

Bud Flagg: Crop farmer, owned land on eastern edge of the tract. Uncle to David Flagg.

Vint Couper: Cousin of John Adams Kimball. Orchardist, benefactor, and beloved infidel. Planted most of the apple trees at Nagog Hill Orchard. Owned land east of the village.

John Morrison: Orchardist and latecomer to the town. Restored the orchards and owned much of the land at the probable village site.

David Flagg: Farmer and landholder east of the village site.

Linda Cantillon: Employee at the local school cafeteria. Halted a million-dollar development project west of the Christian Indian village site.

Sacha Bloch and June Jones: Founding members of the New View co-housing project in Acton, Massachusetts.

Nigel C.W.T. Halstead: Also known as "the Solicitor." English-born American attorney.

Mara Halstead: Former wife of the Solicitor.

Rick Roth: Gamekeeper and caretaker of the Sarah Doublet Forest.

Charles Roth: His father. Laid out the trails at the Sarah Doublet Forest.

Prologue

There was an estate at the top of a hill in the town in which I grew up owned by a man we used to call Old King Cole. The house was a vast brownstone structure with spired turrets and a mean-looking iron fence surrounding it, the type of fence with spear-pointed tips. The grounds, which purportedly had been laid out by the firm of Frederick Law Olmsted, were extensive and unmanaged, with two immense copper beech trees framing a briar-strewn entrance, a small orchard just west of the house, a ruined sunken garden with a frog pond, and many species of exotic trees, including, I was later told, a rare Franklinia.

None of these refinements held any sway for my friends and me. There was once money in the community, but in my time, many of the old houses had fallen into disrepair and the older families had become reclusive and eccentric–I remember the story of one old patriarch who was discovered in his carriage house one night declaiming Spenserian stanzas to his brace of donkeys. In this environment, a tradition of rambling freely over property lines had somehow developed, a custom that would be viewed as trespassing in any other community, but which for us seemed a normal way of life. We would set out sometime in the morning, small bands of us, and roam freely through the town, returning to our camps at dusk, like the great heroes of our childhood–Cochise, Sitting Bull, Rain in the Face, and Crazy Horse.

Of all the properties in the community, of all the woodlots, overgrown backyards, gardens, and frog-haunted swimming pools, King Cole's place held the greatest attraction. For one thing there was a deserted carriage house at the back of the grounds to which we had gained access and used as a hideout. But the larger attraction was that, unlike other landholders in the community, Old King Cole did not seem to appreciate trespassers. Periodically he would emerge from the dark interior of his house to reprimand us—a tottering old man with a cane and a palsied hand. One afternoon he surprised two of us and drove us into a walled corner of his sunken garden. Once he had us trapped, he approached, shuffling, his cane raised ominously above his head, and there, amidst the wild briars and ivies, he delivered a resounding lecture on the nature of title. "My property," he intoned. "My holdings. My kingdom. My nation." Then, advancing a few steps, he pointed southward with his cane. "Your property, your nation. Return to your country. Respect lines of demarcation."

It was a good lecture, but it had the wrong effect. Up to that time, I had no concept of the nature of trespass. Forbidden passage consisted of Old King Cole's land and an even more ominous place in the south of the town, called the Baron's, that was surrounded with a high stucco wall and reportedly guarded by Great Danes. With King Cole's proclamation, the lure of new lands swelled within my heart, offering fresh prospects for adventure and travel. But the one thing that stood out from Cole's diatribe was that the world, which up to that time had seemed to me a wide collective space that invited exploration, was in fact divided and quartered, and guarded, and had evolved into a commodity that could be bought and sold and held. Old King Cole had offered us, in effect, the experience that must have confronted the New England Indians—among others—when first faced with the land tenure system of the Western world back in 1620, the overbearing, all-powerful concept of private ownership of land that overwhelmed the whole American continent and successfully evicted the native use of common land. Why was it that even in his last, clouded years Mr. Cole remembered so clearly the laws of title? More to the point, how did it come to pass that King Cole and his stock could legally surround a patch of wild earth with a spiked fence and drive out intruders? The answer to that question, as with many such grand quests, I found in my own territory.

Years later I ended up living in a town that also had a hilltop estate dominated by an old curmudgeon. This nation was different, and far more rural than that of my childhood, and the lord of the manor was a great raging bull who was more inclined to carry a shotgun than a silver-tipped cane. But the effect of the place—and the draw, I should add—was still strong, and lure of trespass was still deeply seated in my soul.

This same hilltop was the probable site of a seventeenth-century village of Christian Indians. These people, probably members of the Pawtucket tribe, having spotted, they believed, the arrival of a new and powerful deity in their land, converted to Christianity and as a result were granted some sixteen square miles at a place called Nashobah, about thirty-five miles west of Boston. Under the direction of John Eliot, the so-called Apostle to the Indians, the Christian Indians set up a village of pole-framed wooden houses and traditional wigwams, planted apple trees, cleared fields for agriculture, cut their hair, ceased dancing, and settled in to live like Englishmen.

According to the legislative powers of the General Court in Boston, the land, known as Nashobah Plantation, was granted to them outright (never mind the deep irony of the fact that it was their land in the first place). Within the boundaries of the tract, the Indians owned their own houses and property and, with permission of the General Court, were allowed to buy or sell plots of land. But twenty-five years later, during King Philip's War, in what amounted to a prelude to the treatment of the Nisei at the outset of World War II, the inhabitants of Nashobah were rounded up and sent to Deer Island in Boston Harbor, where, over the succeeding winter, many of them succumbed.

After the war, a few of the survivors of this ordeal struggled back to the Nashobah area to live out their time. The last survivor, and the main player in this story, was a powerful woman the English called Sarah Doublet, who died, feeble and blind, in 1736, under the care of two merchants from Concord named Ephraim and Elnathan Jones. By way of payment for her care, Sarah Doublet granted the Joneses the rights to the five hundred acres that she had held, the last remnant of the sixteen-square-mile Nashobah Plantation.

That transfer marked the end of Indian land tenure in this part of the world and the beginning of a new era in land-use history. Sarah and her people would have held their land in common and would have made decisions as to its use communally, by consensus. But in little more than fifty years, this

3

system of holding land in common would be subsumed by the concept of private property. Within another hundred years, this new system would oversweep the entire American continent and replace the idea of the common. It was a uniquely American phenomenon, new even to the conquering English and French.

I knew the general outline of the tract that had been granted to the Christian Indians, but until 1985, when an archaeological student working with old maps and original documents managed to make a good guess, no one was certain where the actual village center was located. Unfortunately, the probable site happened to be owned by the old curmudgeon at the top of the hill, who would not so much as allow archaeologists and curious amateurs to walk on his property, let alone dig the area. The student abandoned her project, state archaeologists lost interest, and research languished. But the more I learned about this area, the more I walked the walls and pored over the scant historical records, the more I became convinced that inscribed there in the old stone rows and wooded slopes was the story of land ownership in the Americas.

That history, no doubt, could be told by examining in detail any plot of land. But at Nashobah, where the English and the Indians—for a while at least—dwelt in a certain harmony, I discovered a splendid mix of sacred lands and profane laws.

By 1736, Sarah's tract was all in private hands, and it remained so until 1988. By 1990, through a curious series of events and coincidences, the tract began a slow, legal evolution back into common land. Two elderly women donated some ninety acres of the original village tract to a local land trust, thus opening up one section of the village site to the public. Then in the mid-1990s another section just to the west came up for development and inspired a small group of people to rise up to save the land as open space. Finally, the core of the place, the sacred geography of Sarah Doublet's final five-hundred-acre tract held by the old curmudgeon, came up for sale.

For several years it was my custom to trespass on the inner sanctum of this piece of private property. But in the autumn of 1995, as a result of new evidence of the spiritual importance of this tract, I began to take my explorations more seriously. Armed with a seventeenth-century map outlining the holdings of the Christian Indians, I set out to walk the boundaries of the ancient property to see what I could see and learn what I could learn.

In the end it turned out to be my most adventurous trespass.

Chapter One

... Know ye that I, Sarah Doublett ... in Consideration of the sum of five hundred pounds in bills of Credit to me paid and for my benefit and advantage secured by Elnathan Jones, Gentn, and Ephraim Jones, Inholder both of said Concord, fully and absolutely, give, grant, bargain, sell aliene, convey and confirm unto them the said Elnathan Jones, and Ephraim Jones, their heirs and assigns forever in equal shares a certain tract of land lying in Littleton, formerly called Nashobah, in said county of Middlesex by estimation five hundred acres....

> *—Land grant from the last known surviving Indian of the*
> *Nashobah Plantation to Elnathan and Ephraim Jones of*
> *Concord, September 24, 1736*

A Certain Tract
of Land

Sometimes in September, when the air is still and the atmosphere is charged with that aura of dormant energy that lingers in certain places long after history has passed them by, I go down to the hill where Sarah Doublet last lived and sit on a wall above Nagog Pond. By general agreement this is as fine a prospect–by day–as any in this area. Sojourners, cyclists, apple pickers, and accidental tourists (the tract is not marked on any maps of the region) happen upon the place and halt in spite of themselves. There is a play of geography here, a sort of remembered country that has worked its way into the Western psyche, a sensation that says, in so many words, "This is the place."

And so the lunchtime runners from the nearby small businesses slow to a walk, cyclists stop to linger on the walls, cars pull over and the passengers debark and stand for a few minutes to admire the view of the orchards silhouetted against the western sky and the fields dropping down to the wooded banks of the pond.

But none of them knows the deeper history of this place, the darker events that transpired in the time when the last band of Pawtucket Indians lived here under the English guns in their squalid little pole-framed houses, dressed in their motley English clothes with their broken dreams of an English god who would carry them, finally, to a better place. None of them knows the story of this Sarah Doublet and the things that happened in this tiny corner of the Western world where nothing of any consequence has happened for four hundred years.

To learn more about this story, I come here at night when the screech owls whinny from the oak hills behind me and the squeaks and whistles of the seabirds on the pond sound out of the stillness. I sit on the pond shore or on the rock walls that interlace the dark wood. Sometimes, with no particular destination in mind, I clamber through the snagged forest and saunter along the edges of the orchards. In effect, I trespass in this place. I *step across,* or *transgress,* in the original sense of the word. By so doing, I go unwelcome onto the private ground of Nashobah, I assault neutral territory. It is, in my opinion, the only way to get to know a place—you have to break through boundaries.

In the autumn of 1675, the English came here with mastiffs and a troop of brigands under the command of a renegade ex-slaver and buccaneer named Samuel Mosely, who had gained a reputation even among soldiers as a man with a taste for excessive violence. His company was a ragtag band of indentured servants, transported criminals, and convicted pirates, released from prison to fight Indians and armed with harquebuses, swords, pikes, and bandoliers.

The Indians of Nashobah Plantation were Christians, and when Mosely summoned them, it is probable that, inasmuch as they were a tractable people, friendly to the English, they would have gathered around him, and when he commanded that they stand in file and march out of the village, it is possible that they lined up and stood in file. It is also possible that they resisted, that they questioned this stranger with his dogs and his pirates. But he was armed and they were not. He was a white Christian, and they, as they had been taught, were but miserable sinners, lowly servants of Christ, docile cows and oxen who labored under the whip of God. And so, it is indeed possible that they went without resistance.

Leaving their goods, their clothes, their axes, knives, and cooking pots behind, leaving also their food, the hard-won larder of a full summer, and facing a cold, snowbound winter when nothing would grow and they would even in the best of times expect to go a month or so subsisting on corn gruel, groundnuts, and a scanty supply of hunted meat, leaving all hope perhaps, and not understanding fully the intent of this man or even his language, they stood in file and, upon command, began their last march.

Sarah Doublet went with them.

So I come here at night, after the runners are gone and the Jamaican pickers leave the orchards, and I go up into the woods above the pond where the village was located and wait for something—I don't know what—the ghost of Mosely, of Sarah, the slicker-snack sound of Ap'cinic, the horned water beast who lived in the depths of the pond in the time when Sarah and her people maintained their garden plots here. I don't know what I'm looking for, but darkness is the place to find it. Night is the last wilderness, the only time when the ancient natural forces that characterize this place make themselves apparent.

Twenty rods south of Nagog Pond in a dank swale of briars there is an anomalous pile of stones. I found it one day after a rain, when the ground had that musky smell of autumn and all the leaves were jewel laden. Whether it was the same pile of stones cited in the early surveys of this tract, I do not know. It's in the right area at least; I am walking on the last tract of Indian-held land in these parts.

Sarah Doublet knew this territory well. She knew the trees, probably even the same species of trees that I know, save for the few that are missing, such as the chestnut, and the few that are added, such as the Norway maple. As a skilled gatherer she would have known most of the plants in this five-hundred-acre tract, would have known where the best berry patches were, would have known where the groundnuts could be gathered and where the little hog peanuts grew. She would have known goldthread and sweet flag and the best spots to gather rushes and reeds for mat making. She was never lost here, she knew the lay of the land, the slopes that I too know, the hillocks, the streams, the pond, the cry of the seabirds that settle on Nagog Pond in autumn before the ice comes in. She knew salamanders and the wood frogs that still drift, legs outspread, in the little temporary ponds here in spring. She would have known the songs of all the birds, the song sparrow, the whitethroat in autumn, the trills of juncos in winter, and no doubt she had heard from her grandmother stories for each, histories, tales of origins and heroes and tricksters, also devils. Hers was a rich, storied culture, thoroughly integrated with the nature of the place in which she lived.

Sarah probably knew of the tract from childhood, but after 1655 she would have been living here. By mid-century there were twenty-five to thirty-five families on the Nashobah Plantation, about the same number of people as would be found in a traditional Eastern Woodland Indian band. The names of some of these individuals appear in the records in other documents related to court cases around the colony. For a while, John Speen, who organized a choir of singing Indians at the Christian Indian village at Natick, identified himself as a member of the Nashobah people. Their teacher or minister was a man named John Thomas Good Man, whose father must have already been a resident of the area, since he had an eel weir at the site on the nearby Beaver Brook before the village was founded. A man called Crooked Robin was a member of the group, and another man named Nepanet who, later in life, married Sarah Doublet. Thomas Waban of Natick claimed some connection with Nashobah, and then there were the women, several of whom were named Sarah by the English. One of these Sarahs was related to John Tahattawan, the son of the founder of the village, the sachem Tahattawan. He and a powerful woman called Squaw Sachem sold Concord to the company of Englishmen who had ventured west to the Grassy Ground River in 1637. More than half the population of the village were women, and there were many anonymous children. But then, they were all anonymous to the English, even in their own time. The English renamed them and even gave them nicknames, some of which were insulting and scurrilous.

Even before there was a Nashobah village, this tract was probably a recognized area, bounded by geographical features and held in common by Sarah's people, not necessarily by law, but by the dictates of cultural tradition. The tract would have been under the control—but not owned by—a powerful male or female leader, and the land would have been held by this person for all the people of Sarah's tribe. Here they hunted and fished, collected nuts and berries, and maintained their planting fields.

With the aid of applied mathematics and instruments to determine contours, lengths, and widths, Sarah's last five-hundred-acre tract has now been measured and bounded and accurately delineated on paper, and divided into a group of some fifteen private lots, owned by private individuals, recorded in legal documents, certified, and filed at a registry of deed on Cambridge Street in Cambridge, Massachusetts. The owners—a term Sarah Doublet would have had trouble comprehending—have the right to exclude nonowners from those lands the recorded deeds have determined to be their property, and they are backed in this matter by a police power and the written laws of the government of the United States of America and of the state of Massachusetts, which is now the name of the section of the continent where

Sarah's people once lived. Because of this, whenever I'm out walking through the woods and fields around here, more often than not I am in violation of the Massachusetts laws that forbid trespass.

Today this area is still generally defined by the two ponds, Nagog Pond on the east and Fort Pond to the west—more or less. There is a swampy little series of wetlands and ponds to the south of the five-hundred-acre tract, and to the north and west are the hills of Nagog Hill Orchards. Just west of Fort Pond, in a section that was part of the original holding of the plantation, there is a high, sweeping hill topped in our time with a red-and-white telephone tower. The town of Acton is just to the south, and the village of Littleton, which now has jurisdiction over this ancient land, is just to the north.

One forbidden zone in the five hundred acres of this former village, the part that surrounds the Nagog Pond, belongs now to the Concord Water Department. No Trespassing signs are posted all over the place. Another private parcel—actually, many separate parcels in this area—belongs to an enigmatic figure named John C. Morrison, the old curmudgeon who does not appreciate intrusion. Another section is divided into long strips that run down to the shores of Fort Pond on the west side of Sarah's tract. And one part, called Sarah Doublet Forest, has been restored to common land by the Littleton Land Trust, which has set aside some ninety acres on the hill between Nagog Pond and Fort Pond in the site where the agricultural lands of the Indian village may have been located. It is the only common land on the whole tract.

Beyond this central five-hundred-acre section, where the village was probably located, is the larger tract of Indian land of Nashobah Plantation, which was broken up after King Philip's War in 1675 as English settlers from the east moved into the region and acquired, by one means or another, the lands formerly held by the Christian Indians. Here, in a region known generally as the Nashobah Valley, the landscape is primarily agricultural, with hayfields and orchards interspersed with low, wet woodlands and sloping, forested hills, dotted now with many roads and houses, a major highway, and a few small computer industries.

One of the prime characteristics of the land in the old Nashobah tract is the presence of water. Besides the two ponds to the east and west of the five-hundred-acre section of Sarah Doublet's land, there is another large pond called Long Lake, where the Littleton town beach is located. Beyond that, but still within the boundaries of the sixteen-square-mile plantation, are three other ponds, woven together by Beaver Brook, which flows northward into the Merrimack River and which, in Indian times, would have allowed free passage for migratory fish such as eels and herring.

The geography of the area would have been no small consideration for the native people of the place. The presence of the ponds, the mainstem tributaries, and forested hills with many good berry patches must have made this part of the world a rich hunting and gathering ground for the Indians. In fact, John Eliot said that "Nashope," as he called it, was the main residence of the most powerful sachem in the region, the man called Tahattawan.

But there is more to the place than resources. Just to the northwest of Nagog Pond there is a low, otherwise unremarkable hill. According to Sarah's people, that hill was hollow and the four winds were pent up inside. Periodically they would attempt to escape, and at these times the bright day would darken, rain would threaten, thunder would cross the sky, and terrible roarings and growls and rumbles would issue forth from within the hill. Then the very earth would shudder, massive rocks would shift from their beds, trees would sway and creak, and were it not for the intercession of their shamans, the earth might have cracked open and revealed the dark, boiling innards.

Later, in the seventeenth century, Daniel Gookin, Eliot's right-hand man, one of the chroniclers of this part of the world, and a famous protector, defender, and spokesperson for Indians, wrote that the local English believed they could hear cannon fire coming from inside the hill, as if an army were trapped there.

Eliot writes that while the Indians were living at Nashobah, there was a great eruption out of the earth in this spot, which left, as he says, a vast "Hiatus" or cleft in the earth and created some immense "Cavities" under the rocks. He claimed that a loud humming noise emerged from the earth beneath the village. He writes about the event as if it were common knowledge, even in London, but says that the place of Nashobah was healthful in all other aspects and that there were some "Goodly Christians" there and others among them "looking that way."

On May 10, 1994, one of the residents in the Nashobah area heard an immense freight train cross the sky above his house. It began in the distance, as trains do, and then crossed a few feet over the roof and disappeared over the hills to the west. Dishes rattled on the shelves, a wall cracked, and then, having failed to escape, the four winds subsided once more.

We, of course, have a different explanation for these internal rumblings. Records maintained by the Boston College Weston Observatory indicate that the Nashobah village site is an epicenter for earthquakes. The first recorded event occurred in December 1668. But Indian legend and contemporary English accounts suggest that there were thunderings and shudders long before and after this date that went unrecorded.

From the west end of Nagog Pond, the land rises in a series of sharp hills, one of them crowned by a granite outcropping topped by a stone wall. Viewed from below, the rock face and wall look very much like a human construction, a ruined crusader castle, complete with crenellated battlements. In fact, the concept of a sort of human geology here is very strong. Some even claim that the landscape around Nashobah—the hills, the standing boulders, the stone walls, and the narrow valleys—is one vast human construct.

Contemporary Native Americans, amateur archaeologists, New Age spirits, mappers, and—privately—a few professional archaeologists believe there is some obscure continuity to this tract that predates the Christian Indian village by as many as two or three thousand years and may continue into our time. Historically—actually, prehistorically—the presence of water, the rocky outcroppings, and, most especially, the thundering hill all would mark this as an important Native American site. Any one of these features would have represented a threshold for native people, a point at which one could cross from the everyday world of temporal events into the spirit world. Some students of this tract believe that there are burial sites in the area, small family groups, perhaps even the grave of Sarah Doublet herself. In fact, as recently as the mid-nineteenth century there are records of Indians returning to what were known as the "islands," where they would camp for a few weeks, to visit, it was said, the ancient burial sites of their ancestors.

Just southeast of the rock outcropping is a low, winding ridge that various imaginative observers see as the form of a serpent. At its tail is a pile of stones that, from a certain angle, could be perceived as a turtle, and not just any turtle, but the world turtle, the great primordial being that, according to certain Native American traditions, rose from the world ocean, bearing upon its back the protoplanet called Earth, future home of a great island, Turtle Island, home of the native people of the latterly named American continent. At the other end of the serpent mound, at the head, rising above an intermittent stream, is a small haystack-shaped hillock with a wooded swamp at its base, surrounded by a bowl-like ridge.

One of my sources on the mythic landscape of this part of the world envisions an elaborate earthworks at Nashobah built, she says, by the "first people," a sort of race of Titans she believes lived here in the time before the

Indians came. Others believe the stone rows that interlace the woods, the piles of rocks, and the small stacks of fallen brush and trees are in fact the work of descendants of Sarah's own people and an indication of a larger spiritual geography of this tract. They argue that Nashobah, with its thundering hill, is the center of a larger configuration, the very heart of an immense placement of stoneworks and earthworks that interconnect the land in this region with the sky and the waters.

Someone, or some group, I don't know who, keeps coming to this area to leave little signs of their visits, as if to say subtly, we were here, or as some observers of this phenomenon would have it, we are *still* here. On top of the mound at the head of the serpent, a site that one of my informants claims is highly charged with spiritual energy, I once found a sort of miniature tepee of small sticks constructed, interesting enough, just below a flowering witch hazel tree. Once I even found a small stone circle in the area, a sort of miniature Stonehenge. Indians did not construct henges in the traditional Celtic manner. The latter-day Indians, of which there are many in this area, tend to leave more subtle signs of their visitation, such as brush piles or, as in one section of Nashobah, an offering of pine cones. But the presence of this model henge, complete with Sarsen stones, suggests that neo-pagans, of which there are also many in the neighborhood, are coming to the site as well.

According to the geologists, professional archaeologists, historians, and local people with whom I have walked this land, none of the seemingly human constructs of the tract is in any way mysterious. The serpent mound is a small esker created by the glacier. Indians never moved stones for fear of disturbing the stone spirits, I was told by one archaeologist. They didn't leave brush donations. The walls are the work of Yankee farmers clearing fields—which is certainly true for most of the walls in the tract—and the stone piles are the work of early surveyors or farmers clearing sheep pastures. The great standing stones in the tract, like the serpent mound, are the work of the greatest artificer of landscape in this region that ever existed, the late great glacier.

But the myth endures. In 1989, two amateur archaeologists, James W. Mavor, Jr. and Byron E. Dix, published a book titled *Manitou, The Sacred Landscapes of New England's Native Civilization,* which has become a sort of bible in New England for people interested in sacred landscapes. *Manitou* has a whole chapter on the site at Nashobah. According to Mavor and Dix, Nashobah is one of the most significant and continuously used tracts in New

England and contemporary practitioners of the ancient religion of the Algonquians continue to "honor" the site.

No professional archaeologist I have ever talked with accepts this view. "It's all bunk," said one. "People see what they want to see, when they need to see it."

The debate over this mere five hundred acres flares up periodically in the regional press, either because of its spiritual associations or because of the threat of development. But one thing that most are able to agree on—even the potential developers—is that the configuration of the landscape in this area is pleasing to the eye. Drive down the present day Nagog Hill Road, or the Nashobah Road, and you will pass through the apple orchards that have characterized this tract of land ever since the Christian Indians of Nashobah Village first set out their trees in the 1650s. From the crest of the hill the road drops steeply, with the orchards to the west, beyond them the wooded ridges, and then on the lower slopes, the wooded hill of Nagog with its serpentine mound, its rocky cliff faces, and the pond called Nagog to the east.

Who really owns this property though? Who really owns any land for that matter? How do you determine where the boundaries lie exactly while you are out walking, and if you happen to cross an imaginary line, one run out and recorded and set on paper and filed in a registry of deeds, what does it matter? The other living things of the tract, which I am informed by legal authorities do not as yet have any rights, freely cross and recross the property lines of this piece of earth. In Sarah's time, I probably could have ranged over this land at will. Now the laws have changed, and what's owned is owned outright and in entirety by its heirs and assigns forever and forever.

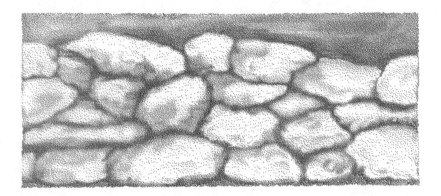

Chapter Two

To have and to houlde, possesse, and enjoy and singuler the aforesaid continent, lands, territories, islands, hereditaments, and precincts, seas, waters,
fishings, with all and manner their commodities, royalties, liberties, prehe-
mynences, and profits that should from thenceforth arise from thence, with
all and singuler their appurtenances, and every parte and parcell theoreof,
unto the saide Councell and their successors and assignes for ever.

—Royal Charter for the Massachusetts Bay Company, 1628

Owners and Outcasts

S arah Doublet, the original "owner" of the tract, had black eyes and a lurid blue image of a bear tattooed on her left cheek. She tied her long black hair in a knot, fastened with a band of silver, and she dressed in a decorated moose-skin skirt and buskins, with a blue shawl over her shoulders and a beaded blue cloth around her waist. Like all the women of her group, she wore thongs of moose hide around her ankles, and in winter, and sometimes also in summer, she greased her skin with bear fat to keep the cold or the insects at bay.

She was born somewhere in the valley of the Grassy Ground River, probably in the late 1640s, and grew up at the knees of her people, hearing the old stories of the time before the English came. Around age fourteen, when she reached menarche, she was taken away to live alone in a small hut apart from the village, where she cooked her own food or was cared for by crones. The males of her band understood the strong medicine inherent in blood, the force of life, so hunters passing near her hut would avert their eyes, or she hers, her power at this time being too strong for daily life, too poisonous. When her period ended, she was brought back among her people as a woman.

At some point during this same time, she would have gone through a ceremony marking her coming of age, but the details of this, as with so much of the richly textured pattern of the lives of her people, are lost. She married young, had at least two children, lost one in a raid, and then having lost two husbands, married again later in her life, this time to the man known to her people as Nepanet and to the English as Tom Doublet. Before she married, while she was still a virgin, she may have worn long bangs that nearly covered her eyes, or a beaded cap with a fringe over her forehead. But when she married, she swept her hair up into an elaborate coiffure with a knot on the top, the remainder left loose or worn in braids with the bright feathers of blue jays and hawks intertwined within the plaits.

Periodically she would paint herself in blues and reds and don a cloak made of bird feathers or a robe of furred mammal pelts, all hung about with heads and clawed feet and the striped tails of raccoons and skunks and made fast with a belt made from the skin of milk snakes and copperheads. She fixed pendants of swansdown or shells in her pierced ears, placed a bird-wing headdress in her hair, and strung herself with shell necklaces and ropes of wampum, and perhaps—all this is conjecture—wore an amulet at her breast, a winged thunderbird or the carved image of Ap'cinic, the horned water monster that lived in the depths of the pond below her village.

After *they* came, after she accepted Christianity, she would have ceased to wear bangles and sparkles and fanciful animal skins, she would have cast aside her bird-wing headdress and her swansdown earrings. She would have become modest, would have lowered her eyes, prayed, sung the strange descant chanting hymns that she and her people would sound out during services. When she was menstruating she would have stayed in the village among the men and no longer would have retreated to the hut in the woods, it being specifically forbidden by the terms of the English contracts that granted her Christianity. According to these terms, she would not weep and tear her hair or smear her face with blood and ash when her fellow tribes-people died, would no longer make a "great noyse by howling" at these times, would not gamble or frolic, would not hold or attend the ceremonial gatherings known as powwows, would not, on pain of death, commit adultery, or pick and eat lice, would not dance, would no longer chant and sing night-long to the slow hypnotic thrum of the great skin drum. She would wear shoes, would dress her hair in a "comely manner as the English do," would, on the Sabbath, attend church and listen for hours to the long harangues and sermons of her ministers, the descriptions of brimstone and fire, and hell, and black-skinned demons with red eyes that fly by night and consume the living flesh of sinners.

By contrast she would have heard of the mercy of Christ the Redeemer, the everlasting God, the very God of very God, in the language of the as-yet-to-be-banned Book of Common Prayer, a god who was begotten, not made, and was of one substance with his Father, God, a Lamb who taketh away the sins of the world, sins of which, by virtue of her birth alone, this Sarah Doublet of Nashobah was guilty.

No matter. Sarah was used to long sermons, harangues, long speeches at her former council meetings. She may even have delivered them herself. She knew the terror that flies by night, the fiery worm, the gnashing devils, and the legends of the tentacled horned monster who would reach up out of the dark waters of Nagog at certain times and draw the entrails of passing villagers down into the depths. She knew the fear of Hobamacho, of night woods, of the screams and howls of her own demons.

She may have been a commoner, an obscure, quiet figure, just one of the fifty to seventy-five people who lived at the Christian Indian village at Nashobah in the fifty years of its existence. She may have been royalty, great-granddaughter of the sachem Tahattawan, who sold Musquetaquid, the Grassy Ground River, to the English. She may have been a leader in her own right, a *saunk* or "Indian queen" as the English phrased it, who could dispense rights and favors and even land to her people. She may have known the great Queen Wetamoo, who fought the English and had a war camp to the west of Nashobah at Mount Wachusett. Sarah may even have heard, in time of war, the deep-throated throb of Wetamoo's drums. But all that would be later, toward the end. And, in any case, little of any of this is known; it is pieced together from the fragments that I have collected about her and about her people.

I spend a lot of time these days sitting on the high wall above the serpent mound, looking down on the sharp little valleys and wetlands of Sarah's domain. I come here in all seasons, find a flat place among the creviced rocks, and sit there thinking out this landscape and the changes that have swept over this region since Sarah's time. Below me, before the village was here, the view must have been the same, the trees, a ruddy swale, the serpent mound, the flickering light off the waters of the pond to the east. Sarah and her people moved in here, cleared off a few tiny planting fields by felling the trees and planting between the stumps, and then moved on when the soils in that area were depleted. After them came the English farmers, who cleared off everything, drew out survey lines, and then fenced their property, first with stumps, then with rock walls. By 1850, when this piece of earth

23

belonged to the United States, it was all sheep pasture and cow pasture. By 1900 it was sheep pasture intermixed with woodlots and cow yards. By 1950 it was scrub, and now it is back to the way it was before Sarah was here, all even-aged trees. The biggest change here is not biological—not yet at least; the land keeps bouncing back from whatever abuse it suffers—the big change is conceptual, a legal attitude toward the place that has changed three times now and, if we are lucky, may change again for the better.

My source on these changing matters of law is an Englishman better known to his American compatriots as the Solicitor. Nigel C.T.W. Halstead tells me that the idea of holding private property in fee simple, that is to say, the absolute ownership of a piece of land that can be bought and sold, is actually a fairly recent development in legal history. The term originated in the English feudal system, when all land belonged ultimately to the Crown. Those who lived on feudal lands were obliged to perform duties, such as military service or farm work, to pay for the right to use the land. Land held with the fewest strings attached became known as *fee simple*. The idea of land as property did not come into full use until the eighteenth century, he says. Before that, in English law at least, what you bought and sold was land held *of* someone; you bought the *right* to live there or the right to use it; you did not actually own the ground. But in the seventeenth century, land came to be seen as an object of quantity, something that, in theory at least, could be sold. Part of this may have had something to do with the fact that the art of mathematical surveying developed about this time and holders of rights of land were able to actually measure land accurately.

By the eighteenth century in Britain, the common rights associated with land—pasturing cattle, for example, or cutting timber or turf—began to give way to a rigid set of regulations based on private, outright ownership of property, and the tradition of the common began to fade. This is the same period as the Acts of Enclosure, when some six million acres of commonly held land—meadows, open fields, and forests—were transferred into private hands by parliamentary approval and were hedged and fenced for private gain.

Here in the new England, even though the idea of the common was still ingrained in the English soul, the concept of the private plot, of each man as lord of his own manor, flourished in the wilderness of the New World. The Jones family who took over the Nashobah property after Sarah died would have assumed the property in its entirety in fee simple, and when they died, since they owned it outright, they could pass it along to their heirs. By contrast, in Sarah's time, and for as many as three to four thousand years before her time, the lands of the tract would have been viewed as a common resource, controlled but not owned by the Pawtucket people. Now, in our time, there are glimmerings of a return to common land.

⚬⚬⚬⚬⚬⚬ ⚬⚬⚬⚬⚬⚬

The legal owner of the core of Sarah's tract under the current system of property law is John C. Morrison, whose name you will see inscribed on the records of the deeds of many of the plots of land in the area and who lives in a white farmhouse on the southeastern slope of Nagog Hill. Here, from his prospect, this Morrison, Black Jack Morrison, as he is sometimes called, or Raging Bull, as my friend the Solicitor calls him, the lord of the manor, holds sway. Sometimes he appears on the porch of his manor house, his great barrel chest bared, his hair cropped close in the manner of a Marine sergeant, his thin legs bearing up the great elephantine body, blue eyes piercing at close range, unfocused, as he surveys his holdings: the sweep of the fields down to the pond, the orchard blocks of Macoun, McIntosh, and Northern Spy, the flocks of ducks and geese of which he, John Morrison, Lord of Nashobah, is master.

No one knows the great man, it is said, least of all perhaps those who are closest to him. That is his choice. He bellows at strangers who dare to step into his orchards. He abuses allies and acquaintances, flies into such red rages against his Jamaican pickers that he reduces perfectly strong men to tears. He is a breaker of stallions. He is a hunter of moose and deer and bear, a man of the North Woods and red flannel who will sometimes disappear into lonely hunting camps for weeks at a time, there to brood and drink and kill animals. There are times, it is rumored, when he stalks his land with arms, the great bull neck swelling beneath the collar of his flannel shirt, hunting boots crushing the frozen grass beneath his apple trees, the ice blue eyes searching. Once, the story goes, he shot someone here in the orchard, a trespasser. He was not prosecuted.

One does not prosecute the lord of the manor. The roads passing through the tract are his roads, the geese and ducks he raises on his property are his geese and ducks, the land, *en fin,* is his land. People in the area make a point of avoiding Black Jack Morrison. He is foul-mouthed and abusive, besotted with power, a ranter, a drunk, a violator of local regulations. He proceeds midroad in his golf cart at a leisurely pace, forcing drivers to move slowly behind him in a great slow train. He pulls out onto his roads without watching because they are his roads. He pays his taxes, therefore he owns the 25 roads. By way of penance, a sort of votive candle lit for the town, he serves on the local appeals board. However, he comes to meetings having consumed more than his normal draft of whiskey and shouts down his opponents. You do not disagree with Mr. John C. Morrison. You do not cut deals with Mr. John C. Morrison, you do not negotiate, parley, talk of philosophy, of nuance, shades of meaning, music, art, or poetry.

One friend of mine, a woman named Mara (who is the ex-wife of the Solicitor), is a literary sort who compares Morrison to the twelfth-century Lord of Hautefort, Bertran de Born, who lived in a castle at Périgueux on a hill above his subjects and went into rages and tirades. He negotiated the fall, it is said, of Henry II, and waged war for the sake of fighting. Ezra Pound says he was a stirrer-up of strife, and the great literary avenger, Dante Alighieri, put him down in Hell and sent him out with his severed head held like a lantern before him.

Such is perhaps the fate of all overbearing landholding lords who abuse their subjects and have no use for local land-use laws. The peasants will rise up, cut off your head, take over your estate, and send you down to Hell.

And yet this Bertran de Born was a famous troubadour, Mara says, a poet, a singer of love songs, gentle with women, beloved by his soldiers. And yet this Raging Bull is a clubman, has allies and friends outside the town, and maintains a reluctantly loyal troop of Jamaican pickers who come to his manor year after year to work in his orchards. Women do not speak ill of Raging Bull, he is more gracious among them. He plays the gentleman. Alone at night, he reads books by his fireside and dozes as he surveys the current literature of pomology. He maintains a horse farm for his first wife and is steward of as fine a property as any landholder in the community. He cares about land. He cares about animal welfare and animal conservation. (Never mind that he once hung a marauding dog by its four legs from an apple tree and instructed one of his workers to shoot it. Said worker, to his credit, cut it down and returned it to the owner.) Raging Bull raises ducks and geese. (Never mind that he was so successful in this venture that his small nascent flock of Canada geese has now established itself and become something of a pest to local reservoir managers.) Raging Bull is a conservationist. In 1962, when a local family proposed to construct a modest summer camp at the north end of Fort Pond, Morrison fought against the proposal because it would have placed a massive tennis dome in an unlikely rural location. He had good lawyers, but so did his opposition—a young local attorney named F. Lee Bailey. Morrison lost.

When a developer sought to tear down the forest at the northwest end of Nagog Pond and put up a group of houses, Morrison attempted to salvage the land by buying it himself to stop the development. He was beaten at his own game and the development went ahead, and then, either to turn the screw, or in ignorance, the developer named the new street Sarah Doublet Lane.

❦

It is possible that Sarah Doublet understood in full the language of the deed to which she set her mark on September 24, 1736. She may have spoken a little English. She may even have been literate. Most certainly she would have heard, almost daily, the lessons from John Eliot's translation of the Bible into her native Algonquian dialect. But whatever her language, whatever her lineage, whatever the social interchanges, dealings, or transactions, the deed she signed marked the end of a ten-thousand-year tenure of Indian landholding and the beginning of an English system that has yet to last four hundred years. It also marked the end of a great, hopeful experiment undertaken by John Eliot, the Apostle to the Indians.

"These strangers," the *Awaunaguss,* the newcoming English-mans, first appeared out of the land of the Massachusett tribe in their dark cloaks of wool, heavy leather shoes set with buckles, and leathern hose or knit green stockings with Norwich garters. They cut their hair shoulder-length and wore it in bangs or parted in the middle. They were clean-shaven and sported no decorations, no flamboyant rings or necklaces or jewels, no earrings. Their shirts were of linen or cotton, and they dressed in suits of Hampshire kersey, and some wore red leather doublets that were fastened with hooks and eyes and lined with linen, and were large and flaring at the shoulders and cut to show the linen shirts beneath. And some others, some few, wore waistcoats of green cotton or leather jerkins. On their heads, they wore black Monmouth hats, some having turned back for convenience the wide black brims and fastened them to the side of the crown. Some others wore smaller knit red caps, and they carried fearsome weapons, these *Awaunaguss,* armor that would repel the weak-willed hickory arrows of the native people, breastplates and flat helmets, and shining swords, and pikes, and muskets, bandoliers, powder and shot. These weapons were the great marvel to Sarah and her people. Sometimes in the villages, to impress the sachems and the sagamores, and, most especially, the *pnieses,* or warriors of the tribe, they would fire off their blunderbusses with great thunderous

27

roars that frightened the children and dropped things from afar. But, weapons notwithstanding, they did not come to the Nashobah region for war—not at first at any rate, not until Mosely—they came bearing the good news of Christ the Redeemer. They came as a clear light, the "Sunshine of the Gospel" breaking forth upon the devil-dimmed consciousness of the blind Indians of New England.

In the 1650s, having translated the Bible into Algonquian, Eliot and his associate, Daniel Gookin, set about establishing a series of villages wherein his converts, his "poor blind Indians" as he called them, could live in peace— provided, of course, they cut their hair and prayed to the proper God. They began in 1654 with a small congregation at Natick, just west of Boston, and, having converted the powerful sachem Waban, organized a series of sermons, some of them preached by Waban himself. By the late 1650s they had secured seven villages of Christian Indians, "praying towns" as they were called, where Eliot's "praying Indians" could live in peace and harmony. One of these land tracts, a holding of some sixteen square miles, was located northwest of present-day Concord in a region of fertile uplands and well-watered intervales. The actual village, which probably consisted of a few pole-framed, English-style houses and many traditional Indian wigwams sheathed with bark or skins, was located between two ponds wherein lay "manie good fishes and planting grounds."

Sarah Doublet lived in a turbulent period of Native American history. Between 1630 and 1633 some three thousand English had immigrated to the Massachusetts Bay Colony, part of the Great Migration. Then, in 1633, a smallpox epidemic swept through the already-decimated villages of the Massachusett and Pawtucket Indians. God had purposely sent this plague, according to the Bay Colony's governor, John Winthrop, in order to "Clear our title to this place." While Sarah was growing up, she would have heard stories of these heavily-clad, arms-bearing *Awaunaguss*. She would be subjected to laws unheard of in her histories, would have seen the old ways rooted out, discredited, banned, ridiculed. And as if to reinforce the strangers' power, plagues and sicknesses that the native shamans could not cure swept through whole villages, killing everyone, including the shamans themselves.

Sarah may also have experienced a disappearance of game. English cows and sheep wandered into the unfenced Indian gardens and destroyed the crops; free-ranging pigs rooted out the mast crops the Indians used and destroyed the food plants of the deer and other native herbivores the

Indians hunted. The English had the strange habit of coming into a forest and chopping down virtually all the trees, leaving a wasteland behind. Worst of all, the fur trade had upset the traditional lines of power in the Indian community, and when the fur-bearing mammals began to disappear, the new Indian leaders began to sell the other resource so coveted by the English, the Indian-held lands. As a result Sarah would have heard of evictions from traditional planting grounds by means of some bizarre arrangement of markings on paper called writing, backed by the orders of a powerful sachemship near the bay in Boston known as the General Court.

This was powerful medicine, this god of the Christians. He did not kill the English with his plagues. He gave them weapons of destruction. What had she to lose by converting? Furthermore, she had heard the "Glorious Newes" of victory over death. Her great-grandfather, Tahattawan, toward the end of his life had heard, via the prolonged sermons of John Eliot (delivered, it is said, in broken, confusing Algonquian that he perhaps only half understood), of this English god. Tahattawan had seen the same disasters, the plagues and powerful medicines of the Englishers. And so he converted.

To demonstrate his submission to the new powerful god they called Christ, he cut his long hair. This was not an insignificant gesture; powerful males of his time arranged their hair into elaborate coiffures or roaches, or pigtails, scalplocks, shaved sections, and tonsures that drew attention to themselves, a vanity that the Puritans abhorred. These English, after all, were the roundheads of Oliver Cromwell's camp, the shorn, virtuous men who despised the vainglorious long hair and powdered wigs of the dashing Anglican and Catholic cavaliers. Laws and regulations in New England were promulgated to discourage the wearing of barbarous hair styles in the fashion of the wild Irish and, worse, the "salvages," or savages, of the forest.

And so Tahattawan cut his hair and ceased his harangues and dancing and his weeping and gnashing of teeth at burial ceremonies and became, as they wished, a docile ox laboring beneath the yoke of Christ. And his great-grandchild Sarah became as a weaned child, a tame cow, submissive, obedient, and gentle. What had she to lose? And what to gain? If nothing else, the new medicine might hold at bay the horned water beast that slept in the waters of Nagog Pond.

Chapter Three

Inanimate objects are sometimes parties in litigation. A ship has legal personality.... The corporation ... is an acceptable adversary and large fortunes ride on its cases....

So it should be as respects valleys, ridges, groves of trees, swampland, or even air that feels the destructive pressures of modern technology and modern life. The river, for example, is the living symbol of all the life it sustains or nourishes–fish, aquatic insects, water ouzels, otter, fisher, deer, elk, bear, and all other animals, including man, who are dependent on it or who enjoy it for its sight, its sound, or its life. The river as plaintiff speaks for the ecological unit of life that is part of it.

–Supreme Court of the United States. No. 70-34 (1972)
Justice William O. Douglas, dissenting

Should Trees
Have Standing?

One October day I was sitting on a stone wall above the orchards meditating on the curious geological formation of the earth in the place—the humped landscape, the sharp little hills, the granite outcroppings and low wooded hills, low-lying swamps, and thickets. I had attempted to walk the northern boundaries of the Indian village that day, but as usual, I had gotten lost. More accurately, I had allowed myself to get lost. It's part of my exploration of this world, a way of finding out where you are exactly.

I had started from a small pond up on the northeastern boundary and then threaded through backyards and fields, trying to trace the angles and turnings of the old surveys of the area. But no luck, so I worked my way toward the center of the tract, which is the orchard of old man Morrison. I crossed the orchard and found a sunny spot on a wall above the pond and sat down to think.

I grew up in a household that eschewed ownership of land. One of the dictums of my old father was his singular vow never to own another blade of grass. This came to pass partly from an ethical stance—what right do we have to "own" earth—and partly from the realities associated with owning too much. My family once held large tracts of farmland on the Eastern Shore of Maryland, most of which they worked by leasing out to tenants. For who knows what psychological reasons, shortly after he graduated from college my father attempted to excuse himself from this system and went as far away as possible from the Eastern Shore of Maryland. He took a teaching position in Shanghai and stayed there for three years. Nevertheless, the transfer of privately held lands being what it is under the American legal system, upon the death of a landholding uncle, my father inherited a three-hundred-acre farm, complete with a house, barns, and its own tenant farmer. He came home from China, revisited the old homestead, and after some moral debate, decided to enter the ministry and took a church in the North. He still owned the farm though.

By now it was the height of the Depression and in spite of the fact that he had deserted his roots, since he had a salary, he actually had more money than the landholding gentry in his family, some of whom were so poor they ended up burning their corn crops for heat in their large, chill town houses. My father's property, known locally as the Driver Farm, was one of the best croplands in the area, but, given the economic conditions, he was losing money nonetheless. One year they had to turn under their entire tomato crop because it was not worth transporting to Baltimore to sell. One night, shortly afterward, during a dinner party, so the story goes, my mother answered a long-distance phone call and returned to the table ashen white. The tenant farmer on the Driver Farm had killed his wife and then shot himself.

My father went "down home," as my parents always called the Eastern Shore even after fifty years' absence, and closed down the affairs of the farm. It was all a hideous mess. In a bizarre twist, related to me later, the right hand of Mr. Weir, the hand that held the pistol, had turned black. On his return my father preached a sermon on the tenancy system, on the immorality of land ownership. He read deep metaphors into the message of the Black Hand. Then he sold the Driver Farm for very little money and vowed never again to own another blade of grass.

There are times when, in reaction, I fancy myself a country squire with vast properties, an estate of woods and fields, with a small house and an English garden and extensive grounds, which, attired in English tweeds, I should peruse on fine mornings. But that, given my current financial situation, is an idle dream, and in any case, it emerges out of some mythical, nonexistent England where all the world was right, and God was in His Heaven.

"God," says my English friend, the atheistic Solicitor, "if he is anywhere—which I very much doubt—is certainly not in the courts of law. And if he is, he is inordinately fond of dirty tricks and has no respect for truth, nor justice. Our work is to trick him."

This so-called solicitor, Nigel Halstead, has his own story. He is one of those individuals, not uncommon in our time, who straddles two cultures. Although he holds a place in the Massachusetts bar, until recently he was a citizen of England. He was born at Harrow not long after, as he explains, the "nasties" were defeated at Berlin. His parents, having only one offspring and wishing to honor his uncles, bestowed upon this happy child the names of all the uncles rather than offend one by omission. In the 1970s, having misspent his youth on the Portobello Road, the island of Formentera, and other popular retreats, Nigel Charles Thomas William Halstead fled to America, married, returned to college, reformed, and went on to take an American law degree. Now he is an established attorney who periodically takes on pro bono cases having to do with the environment. Years ago he departed from Nashobah and, having divorced himself from his family and his life as a good burgher, now holds sway from a local watering hole in Cambridge with a Celtic ambiance, where, because he once represented the owners in a small case, he has become an institution–Nigel C.T.W. Halstead, Esq., the Solicitor for the Druid.

Out of court (and within court, no doubt) the Solicitor has taken on a persona. He dresses in dark wool suits with pressed white cotton shirts, and he favors colorful suspenders and florid ties and combs his graying locks straight back from his forehead. On special occasions, he dons waistcoats and bow ties. Years in the pubs have given him a comfortable middle and a twinkle in his sea blue eye. Half of what he says is set up for a point he's trying to make, half he actually believes. The Solicitor is a good arguer, but it's hard to know what he thinks.

⚬⟞⟝⚬

While I was daydreaming at the wall, I saw another denizen of this place, a lank, ragged figure working his way slowly up the slope through the spidery trees, still hung with heavy red fruit. I knew this man even from a distance. It was my friend Mr. Vint Couper, who owns some sixty acres of hayfields and apple trees on the other side of the hills from the Nashobah site.

Mr. Couper, who is in his mid-eighties, is known locally as the barefoot farmer and is much loved in this community; you often see him around town, and those who drive up and down the Great Road past his farm stand see him at work by the side of the road. He's hard to miss. Mr. Couper is not a man who concerns himself overly with fashion. He dresses in a piebald, motley collection of shirts and pants, sometimes worn one on top of the other, that he appears to have salvaged from the dumpster at the town supermarket parking lot. His great claim is that he has not worn shoes since 1922, when, for reasons he does not entirely understand, his feet began to feel cramped. He secures his pants with a rope strung over one shoulder and fastened at the top button of his fly, like half a suspender. In order to keep his cuffs out of the mud, he cuts his trousers off above the ankles, about halfway up his shin. When the weather closes in, he dons a conglomeration of flannel shirts, sweaters, winter coats, and a woman's wool hat. In effect, anyone who sees him and does not know the man would presume he was one of the homeless unfortunates who periodically wander up and down the Great Road for a few months during warm weather and then disappear. In fact, Mr. Couper is a member of one of the oldest land-holding clans in the town.

"Seen Dennis?" he wanted to know, after the proper greetings.

Dennis Marchand, the Jamaican crew boss from Clarendon, has come back year after year to work for Morrison. He was one of my best sources for stories about the old man until, for reasons I never quite understood, he stopped telling stories.

"I left a row of fruit baskets out here last year," Mr. Couper said, "and Morrison told me Dennis put them somewhere. 'Go find him,' he says. 'Where is he?' says I. 'Don't know,' says he. End of commentary."

In recent years a sort of quiet feud had developed between these two old men, originating, it seems, years back when Mr. Couper caught Morrison in an embarrassing moment with a younger woman under the apple trees.

I didn't want to miss out on an opportunity to learn more about this story since I had never managed to get the details clear, and because we were now sitting at the very scene of the crime, I managed to bring the subject up.

Mr. Couper happily told me the story again, as if for the first time—how he was driving the tractor to a remote section of the orchards and how he passed our man in an amorous embrace, and how, inasmuch as Mr. Couper was a gentleman and not some kind of Peeping Tom, he averted his eyes and carried on.

"He knows I know, see," said Mr. Couper. "He saw me see him, and I saw him, and I passed on with not a word. Would have kept my mouth shut about all this. What do I care about his private life anyway. The next thing I know people around this town are saying *I'm* a sexual maniac. Well, I'm not a sexual maniac. But I know who started that rumor."

I have been collecting stories around this town for twenty years, and never have I ever heard anyone suggest that Mr. Couper was a sexual maniac. In fact, the one thing that you hear about Couper is that, in spite of his couture, he is a perfect gentleman. Eighty-four years old, limping around with his bare feet and ragged clothes—maybe it was wishful thinking. Mr.

Couper has never married (at least "not yet," as he always marks on his tax forms), and has spent the last twenty-five years living in squalor in the cellar of his own house, in a room he proudly calls the Snake Pit.

Story or no, the orchards are as fine a place for a springtime tryst as any, and I sometimes see lovers strolling here. Once, down by the Nagog Pond shores, I even found a remembrance of a long-lost love. Someone had pinned a little paper heart with some writing on it to a tree trunk overhanging the water. "Here, beside the still waters," the inscription read, "two lovers loved."

Dennis was nowhere around. He never was an easy man to find except at noonday, when he would come into his quarters for lunch, which is where I would usually corner him. He is the headman, or at least the longest-serving member of a crew of some eight to ten Jamaicans who come up every season to Morrison's orchards and are housed in a barn at the crossroads on the top of the hill. He is a fine-boned man, wiry, with yellow in the whites of his eyes and a bright golden star cut into his left front tooth. Even in summer he wears dark flannel shirts and heavy dark pants tucked into rubber Wellingtons. He often works in the orchards with a well-sharpened machete in his belt, which he uses, as he would in Jamaica, to cut grass and brush. But the gleaming blade, and his habit of slicking it over a sharpening stone in idle moments, suggest a certain stalking menace. At the very least it must offer some protection to the orchard cash box that is left unguarded by the side of the road at Morrison's pick-your-own-fruit operation.

In fact, Dennis is a man of peace, as are all the workers at Nagog Hill Orchards. At home, if you can believe his stories, he is more to be feared. Once, he told me, he returned to Clarendon to find that his woman friend had been stolen by another man. Dennis hunted him down and in one version "fixed" him—whatever that means.

Dennis has a brother, Jeffers, a less visible man who does his job and keeps out of the way of the old man as best he can. Another regular at the orchard is a quiet worker who always wears a knowing smile named Carleton, and there is another, a huge man the whites call Aska. I once saw a document that listed this man's given name; it was Oscar, not "Aska"—which says something about the confusions of commands and mistakes that sometimes take place at Nagog Hill Orchards because of regional accents, the local Massachusetts accent as well as the Jamaican. Once when I was telling Dennis that I was interested in writing about the law in this area and how it applies to the land and the orchards, he said I should go talk to a local man named Wayne Wade. "He is a great liar," Dennis said.

"Why would I want to talk to a liar?" I asked. "I'm trying to write about the truth, the law."

"That's right, mon, he knows all about de la. He a good liar."

Dennis, who as headman has the right to complain more about Morrison than anyone else, will admit that the old man has his good qualities. Among other things, he pays well. Minimum wage for workers floated around $4.50 per hour and big corporations submitted testimony in Congress against any increase. But Black Jack Morrison paid his workers, whom he flies to and fro from Jamaica at his own expense, $7.50 an hour, or thereabouts. He then sends his accountants with them to the bank on payday to make certain the money, or a certain portion of it, is invested properly so that when they return in late autumn to Jamaica, they will not go back with empty pockets. More to the point, perhaps, these men return year after year.

The sun was warm on the wall. It was now midafternoon, I had been up since early morning, and after Mr. Couper wandered away to look for Dennis, I settled in a nook in the wall and started to doze off. It was a good spot for a rest, the October sun was warm on my face, and I only woke up because it swept west over the hills and the air grew cooler. No one was around. It seemed strangely quiet in the orchard, as if the world had ended while I slept.

Just before I got up to leave, below me on the road, I saw *him.* The great bull. He emerged from the machine shed where the Jamaican workers were housed, mounted his golf cart as one would a chariot, and steamed off down the middle of Nagog Hill Road on some errand. He was hunched over the wheel, head tilted forward, chin jutting, the great tight arms crooked, dressed as always in a flannel shirt and ragged trousers–like Mr. Couper and many of the other farmers in these parts, he was not a man who concerned himself with style and, in spite of the fact that he could buy and sell the town, was sometimes taken for the local bum.

I stood up and surreptitiously vaulted over the wall onto the sanctuary of the common land of the Sarah Doublet Forest. In spite of my interest in his property, I confess I have never met this Raging Bull, and from what I have heard, I'd rather not.

On the other side of the wall, I came face to face with another demon of this piece of earth, a figure out of *Spiritus Mundi* with a glistening black snout and a huge barrel-like belly that nearly scraped the ground. It was the caretaker's pet Vietnamese potbellied pig. Rick Roth, who works part-time on the land trust property and lives just above the Morrison orchards, keeps a couple of hogs, one of which is this vast oversized hybrid he calls Homer. Said pig is an artful dodger and is forever escaping his pen, which is why I happened to come face to face with him that day.

On one of his earlier forays, as animals will do, he crossed the legal boundaries of the Morrison land, discovered below the hill the fine gardens of John Morrison, and proceeded thereto in order to feed. When he was finished his repast, he had consumed the better part of Morrison's carrot crop, plus a fine selection of new potatoes, topped off with a row of lettuce. Having dined sufficiently, he retired to his pen to sleep off his meal. Unfortunately, Morrison saw his great black back retreating into the brush and stormed up the hill to berate the pig of Rick Roth. His tirade grew to include the wife of Rick Roth, the children of Rick Roth, the ancestors of Rick Roth, the dogs and cats, and, most especially, said pig. If he returns, Morrison said in so many words, he will be spitted and roasted and consumed with relish.

Homer did return, was not caught nor roasted, and Morrison came again to the house to complain, this time with only slightly less ferocity. His style of visitation was unique. He would pull up in a car and beep until some adult human being emerged. Or he would storm through the orchard in his golf cart and lean on its horn until someone appeared. On his second visit, before complaining, he paid a call on Homer in his pen, and apparently spotted there a certain kindred spirit, a sense of survival perhaps, or a recognition of communion. (Cynics have suggested he saw there some element of himself: a great, hefted beast, fond of eating, skinny legs, and an ability to slash at things. Homer once opened a long gash in the side of one of Roth's prize hunting dogs.)

After that Morrison returned repeatedly to Roth's property, not to visit Rick, I was told, but to visit Homer.

Rick Roth is part of a loosely connected community of people in their twenties and thirties who for whatever reasons have been unable to depart from the territory around Nashobah, where they were born. In keeping with the manorial aspect of the Nashobah lands, Rick is by profession a gamekeeper, a job that is almost unheard of in this country, but is a respected position in rural England. He works for a gentleman hunter who owns a large private reserve in southern New Hampshire modeled on the old royal deer park. No one but the king and his entourage is allowed into the forest, and then only in the proper season, and furthermore only if a hefty annual due is paid to the owner, something like $10,000 a year. The owner often visits England and periodically takes Rick with him to inspect local reserves. He attempts to maintain here in the hinterlands of America, in the brutish tangles of the Monadnock intervales, a remembrance of things gone by. For example, his gamekeeper, who in everyday life favors sweat shirts and jeans and hiking boots, is required to dress in English tweeds whilst on the reserve.

Rick's father, Charles Roth, who laid out the trails at the Sarah Doublet Forest, is one of the old lions of the environmental education movement in this country and knows a thing or two about saving open space. One document refers to him as the father of environmental literacy. For twenty-seven years he was director of the education programs at the nearby Massachusetts Audubon Society, and before that he worked as director of various nature centers in Connecticut. On retirement, or what passes as retirement for old nature men like Roth, he developed plans for state environmental education and worked on the local land trust that preserved the Doublet tract. He also began to research such arcane subjects as the ancestry of Sarah Doublet. He happens to be a descendant, through his mother, of Roger Williams, the seventeenth-century cleric who had an anthropologist's eye for native society and even prepared an early phrase book of their language. Roth, who looks more the part of a seventeenth-century cavalier than a twentieth-century educator, is right out of Franz Hals. He has a van Dyke beard, a mane of yellow-gray hair, and full florid cheeks. I imagine him in seven-league boots with a flagon and lute.

When the house above Morrison became available, Roth and the land trust managed to salvage it, restore it, and then place son Ricky *in situ*, with his small family, to oversee the property.

Not to stretch the analogy, but such situations were common on the English estate: a manor lord, a gamekeeper, a wise thane or overseer who manages the system, common land and private plots, forests set aside as forest, orchards and fields, and the whole of it supported by impoverished, landless serfs who live in housing near the manorial estate.

There is even a court jester at Nashobah, or in this case, a sort of jester chorus, consisting of the troop of young, witty, somewhat intellectual people who, as I said, have yet to leave the town for larger prospects. You will see them on hot summer days at a place called the Landing on the town lake just west of Fort Pond and Nagog. They turn up en masse at birthday parties, weddings, and local events, and are responsible for an elaborate puppet show, intended for children, at the town's annual September fair. One of their party periodically organizes masques in the style of Henry Purcell's *Indian Queen*. One of the most memorable featured a staging, in tableau vivant, of Botticelli's "Madonna and Child." When the curtain drew back on the makeshift stage, there sat, center stage, one of the young women of the group attired (somewhat ironically, given her sexual history) as the Virgin Mary with her nine-month-old baby at her breast. Unfortunately, when the curtain was pulled back, the Baby Jesus, spotting an audience, began babbling and squirming and waving His arms. (Never mind, by the way, that Our Lord happened to be a baby girl—that is beside the point.)

<hr />

To outsiders who happen upon this area, the land just north and east of Sarah's tract is the best-known section. It borders the Great Road and includes the popular ski resort for Boston known as Nashobah Hill. There are still working farms on either side of the road, and throughout the growing season this section offers a classic rural landscape, with old apple trees and sagging barns and greening hayfields rolling off to wooded ridges. But all this is owned by people whose only wealth is in the land. They have no retirement funds, no formal savings systems, no health insurance. It is these properties that the hard-working people in the community who serve on the town boards, such as the conservation commission, and those equally hard-working people who make up the town's nonprofit land trust, which buys open space for the benefit of the public, worry most about. This singular stretch, coupled with another section on the west side of the town that used to be called Scratch Flat, creates the defining character of the community. Here, lined up like the fast-food joints of other towns throughout America, are makeshift stands where the local farmers sell their produce.

None of these lands are currently protected, they are all private, and some of their owners and immediate families, who in their own way have been great preservers of open space (outside of California and the Midwest one does not farm in America without at least some appreciation for land), are not necessarily planning to preserve the lands after they die, or need cash for retirement. Furthermore, some of them have strong suspicions of government in all forms.

The closest farm to the five-hundred-acre tract of the Doublet land is held by Vint Couper, who has some sixty acres. Across the Great Road to the northeast is a series of hayfields owned by a local farm family who survives in part by selling pumpkins and cutting firewood, a common source of heat in the rural households in the area. Just to the north of this property is a farm stand known as Nashobaside, run by a man named Bud Flagg. 41

Mr. Flagg clips his hair, which is still jet black in spite of his age, in a flat-top crew cut. He has clear blue eyes and the kind of glasses with wire rims on the bottom and solid black plastic on the top that went out of fashion around 1960. He used to smoke a pipe with an intertwined metallic stem meant to cool the smoke, a fashion also in vogue in the late 1950s. In short, he is a man who got stuck in time, not an uncommon phenomenon for those who spend their days working in the fields.

This failure to catch up with the times is a characteristic shared by Mr. Couper as well. I once asked him if he knew about computers. "Well, I have heard about the instruments," he said. "But I can't say I have actually *seen* one."

I once spotted these two struggling with the instructions at an automatic teller machine at the town bank while a long line of restless computer workers waited. Each new instruction from the machine introduced yet another mystery which the two old men had to discuss at length.

"Your *name*," said Mr. Couper with authority, "is not your *code*. That's meant to be a secret, see. Don't tell me that again." Mr. Flagg pushed some more buttons.

"Now you've poked the wrong one, Bud," he said. "You got to poke in the *code*. It doesn't care any about your name."

The youngest of this set of landholders is Bud Flagg's nephew, David, who specializes in herb growing and has, some would say, an unhealthy suspicion of government controls. He lives in the old family homestead with his mother, an eighty-year-old who used to putter around her flower beds in a print shirtwaist dress, wore classic silver-framed granny glasses, and had white hair swept up in an 1890s-style bun. David Flagg has been a vocal protector of private property rights at local town meetings, and would object vociferously even at the *suggestion* that his right to do with his property as he wished might be curtailed by government injunction. "You own land," he says, "and you own with it a bundle of rights, just like a bundle of sticks, guaranteed to you by the Constitution. Now, one by one, those sticks, those God-given rights, are being taken away by government regulations."

In spite of his stance, neither he nor his mother, the legal owner, has sold any of the fifty-seven-acre farm to private developers, and in fact David Flagg spends a lot of time and energy worrying about the fate of the local farms.

Farmer Flagg, as he calls himself, has successfully assumed the character of the jolly farmer. He's a round man with a full face and rosy cheeks and a goatee, and in summer he sports a wide-brimmed straw hat, green twill pants, and a blue flannel shirt open at the neck. But there are not a few anomalies about his farm and character. One is that he is a well-read college graduate who will just as likely draw you into a discussion of Russian history as the weather. Furthermore, his farm, which is a clutter of discarded plant trays, pots, machinery, and makeshift greenhouses, is periodically flooded with the strains of Mozart and the cries of the exotic fowl he collects as a hobby.

He and his family are newcomers to the town, having bought their property from the Kimball clan in 1830. The Kimballs in turn got their lands in the late 1600s, even before Sarah Doublet signed over the rights of Nashobah to the Joneses. That original Kimball tract is now split between Mr. Couper, his ninety-year-old cousin, John Adams Kimball, and the Flagg holdings on the north side of the Great Road.

If all this sounds vaguely feudalistic—one or two families holding large tracts of land—that is hardly the case. These are men and women who have the good earth encased in their hands, not gentlemen farmers. But it is a doomed occupation, a doomed land they work. All these people, save David Flagg, who is in his fifties, are over eighty. Farming in New England is not exactly a lucrative business, and surrounding these rural tracts, east, west, north, and south, are the tentacles of urban centers, grasping at the open spaces beyond their pulpy bodies. As Farmer Flagg points out, it's just a question of time before they curl around the farmlands of Nashobah.

You have but to walk a quarter of a mile up from the low swamps of the old Nashobah village to a high hill just west of the tract to see the problem. Here there is another prime piece of real estate, consisting of some one hundred acres in two tracts of rolling high hayfields that drop down to the wooded, as yet undeveloped, eastern banks of Long Lake. A few years ago this property was sold to a developer who planned to place there a fashionable shoreline development of some sixty-five large, expensive dwellings, with a swimming beach, to be sold to the rising magnates of the local computer companies at three hundred to four hundred thousand dollars a clip. And why not? This was private land, owned in fee simple and, with a signature, could legally be transformed from farmland and woods into whatever use— within the zoning codes of the town—that the private owner wished.

But, as my friend the Solicitor might argue, what about the rights of the grassland birds who still breed in the hayfields in blissful ignorance of American property law? What about the rights of the thousands of people who, passing over the hills of Nagog Hill Orchards, happen upon this rising sweep of hills and enjoy the view. What about the fact that open spaces throughout this community, throughout this whole continent, in fact, are disappearing beneath impermeable structures such as houses and parking lots, which evict, for all intents and purposes, and for all time, the complex living organisms that have endured there ever since the ice sheets withdrew and exposed the northern half of the American continent. Have these living things no rights? Should not trees, rocks, mountains, hills have as much legal standing as, say, a ship or a corporation?

This question, according to the Solicitor, was first addressed in 1972 in an essay by Christopher Stone, a professor of law at the University of Southern California. The argument, which proposes that natural objects should be granted legal status, actually has ancient roots. Roman law, Eastern law, and even medieval church law granted certain rights to animals, for example. The Christian church would sometimes try renegade donkeys for their sins before administering punishments. (In one such case, a donkey was sentenced to prison. He was retired from his strenuous workday and was allowed to stand in a stable, where he was fed regularly until he died of natural causes.)

In American legal thought, the idea of the rights of natural entities began with an essay by Aldo Leopold called "The Land Ethic," which was first published in 1949. Leopold used the analogy of Odysseus's treatment of his disloyal slave girls. When he got back from Troy and discovered their transgressions, he had them all hung. They were his property, he had the legal right to dispose of them as he saw fit.

Stone traced the development of this idea of private ownership and suggested that the granting of rights to things that formerly held no rights, such as slave girls, has been an evolution. Women, old people, children, slaves of course, and other entities, such as corporations and ships, originally had no legal standing but have acquired rights through a slow accretion over the course of civilization. So, presuming the wisdom of law to be an evolution, why should not trees have legal standing? Who else would speak for them but the law?

It was a very dangerous question.

It was also a question so simple in its directness that a resident who works in the local school cafeteria and used to walk in the hilly tract west of Nashobah came up with the idea on her own when she learned that the land where she would take her outings was slated for development.

Linda Cantillon wears white tennis shoes, dresses in jeans and cotton shirts, and lives in a small house on a small street in a small town with two small children—not the type of person who normally strikes fear into the hearts of land developers and corporate magnates. But as it used to be said in the halls of Congress even as early as the 1960s, "Beware of women in tennis shoes."

It was, for example, a group of women in sneakers who started the 1971 Mineral King Valley case, in which Walt Disney Enterprises sought to transform a wilderness area of the California Sierra Nevada Mountains into an immense complex of motels and recreational facilities. That case went all the way to the Supreme Court, cost the Disney corporation untold millions in court fees, and generated a famous dissent by the great legal defender of wilderness, Justice William O. Douglas, who argued, citing Stone, that environmental objects should have some means to sue for their own protection.

Linda Cantillon started out as a simple lover of wildflowers who would take her children down her small street, cross an old wooden bridge over a slough, and head through the woods to the high fields, where, each in its season, grew Queen Anne's lace and hawkweed, Indian paintbrush, yarrow, mullein, quaker-ladies, and cinquefoil. Here, also in their appointed seasons, little bubbling flocks of bobolinks would rise and settle from the high summer grasses, meadowlarks would call, and she would see the flights of small brown, unidentifiable sparrows and, high above, in the summer air, the specks of swallows and swifts darting after insects amidst the rising clouds. It was a consummate nineteenth-century rural idyll. Winslow Homer, were he with us now, would have painted Linda in her straw hat among the long grasses with her attendant children.

Almost daily in the proper seasons, sometimes accompanied by her mother, she would walk the land, never dreaming for a moment that she was breaking the immutable laws of trespass. Her mother taught her a love of wildflowers and the subsequent art of trespassing. While she was still a girl, the two of them would roll around the countryside in their old station wagon with shovels and bushel baskets in the back, looking for likely patches of flowers. They were like conservation archaeologists in their quest. Whenever they spotted an area about to be destroyed by development—not a difficult quest in eastern Massachusetts, it must be said—they would, sometimes by cover of night or early morning, descend on the place to rescue the flowers and vines and small trees that her mother would then transplant to her yard. Linda has carried on this tradition. Her yard is an apparent mass of weeds, interspersed with freshly dug beds of weeds, interlined with borders of weeds. In any upscale community her property would be considered a disgrace, except that, as Emerson said, a weed is but a flower out of place. Throughout the growing season this yard flourishes with a profusion of wildflowers that struggle against one another to gain ascendancy. Once her flowers are planted, it's every weed for itself. She neither cultivates nor lays down mulch. "I hate mulch," she says. "I don't want this place looking like a bank."

The last thing on earth any passing stranger would say about Linda Cantillon's house is that it looks like a bank.

One day in the high fields above her house, she saw an ominous sign.

"These guys, right? These three jerks in short-sleeved shirts and neckties out in the fields with their construction boots and clipboards, looking down a hole. Made me nervous. I've seen this before with my mother. Developers."

This was the first time she understood that the fields and woods where she had been walking for three or four years were not public land. "I always though this was the conservation land," she says. "Next thing I know they're telling me that there are sixty-five houses going in here, a beach down at the lake, golf course, mini-sewage-treatment-plant, these big, fancy four-hundred-thousand dollar houses. No way."

And so, armed with nothing more than passion, an innocent at play in the fields of commerce, devoid of political experience, devoid of any knowledge of the intricacies of American law, this Joan of Nashobah set out to transform a tract of land that had been held as private property ever since Sarah Doublet signed her deed back into common land.

❦

This was not the only fight that was starting up at the time. Two other developments near Nashobah were proposed that year. One was located on Beaver Brook, just downstream from the spot where Sarah's husband Tom maintained a fish weir. This too was to be a large traditional development with expensive manorial-style houses, roads, driveways, lawns—and it too was to be constructed on ecologically sensitive land bordering Beaver Brook, officially cited as one of the five richest areas for wildlife in the state.

As is often the case, when the nearby residents heard the news, they banded together, held committee meetings, and began plotting ways to stop the development. It was the standard reaction, I was told by an agent for the developer. Mister Developer himself was used to negative response by now, the not-in-my-backyard philosophy that he would meet with every new proposal. What for the neighbors became an intense, impassioned fight, with news campaigns and interminable night phonings and weekend meetings and not a few tears and frustrations, was business as usual for the developer.

The other adventure that was shaping up just south of this property was a proposal to create, on a hilly section just southwest of an open field, a development of some twenty-five houses. But, unlike the other two proposals, this development would use only half the land for housing and leave the rest in common ground. The dwellings were to be nestled together into the hills in such a way that they sat in some cases no more than a few yards from one another. In the center of the development there was to be another structure, a central common hall, a little like an Anglo-Saxon mead hall or an Iroquois longhouse, with sitting rooms, a kitchen, and a large dining room. The idea was that the people living in the small houses surrounding the central hall would eat together each evening—if they so desired. Food would be prepared on a rotating basis by the residents. Land in this development would not be owned privately, save for a small patch underneath each house. The rest was common land, belonging to all the people in the "village" in equal shares. All decisions as to the use of the land, and in fact all decisions affecting the group, were to be made by consensus. The idea was very similar to the social structure of the Indian village at Nashobah, or even a tenth- or eleventh-century manorial estate, save that in this case the sachem, or the lord of the estate, would be replaced by a village committee. It was a downright anachronistic idea, and it went against everything that was American. More to the point, in order to be constructed in the first place, it would have to break almost every zoning and building code in the town of Acton, where it was to be located.

But such is now the changing landscape of America. The postwar vision of sprawled-out tract houses, each with its own private lot, continues as if there were no boundaries, no limits set by unbuildable land or lack of water. Meanwhile, in small pockets, mainly on the east and west coasts, the older model of housing concentrated within commonly held land is slowly regaining acceptance.

All three of these systems—the idea of co-housing, the creation of public open space, and, of course, the traditional sprawled-out development—were in process around the old Nashobah tract the year I spent exploring the area.

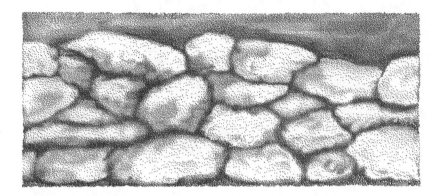

Chapter Four

In answer to the petition of Mr. Jno. Eliot on behalf of severall Indians the Court grants viz; librty for the inhabitants of Nashop and to the inhabitants of Ogkoontiquonkames and also the inhabitants of Hasnemesucuchoth to erect severall injan townes in the places propounded with convenient accomodation to each provided they prejudice not any former graunts, nor shall they dispose of it without leave first had and obtained from this Court.

–Order of the General Court granting land tracts
to Christian Indians, 1651

The Cords of Christ's Tent

Nagog Pond is a good place to look for birds. There is no development on three sides of the pond, there is a good mix of woods and orchards, and for some reason, gulls and other seabirds seem to collect there even though it is thirty-five miles inland. Sometimes I pull my car off the north side of the pond and trespass along the banks, to see what birds are there.

I was driving up to see what Dennis was doing one November day and pulled over to take a walk first, leaving my car beside the road. I had been told that the children of some people who lived in a house close to the shore, in a section as yet not taken over by the Concord Water Department, had found some arrowheads in the area, and I thought, poking along the shore, I might turn one up. I never did, but it was a pleasant, if soggy, walk. The only problem was that when I got back to my car, two policemen were standing in the road staring at it.

"This yours?" one of them asked.

I admitted that it was. They wanted to know what I was doing.

"Just a little bird walk," I said.

"Bird walk," one of them repeated cynically.

I decided if I was going to get out of this one, I was going to have to play it to the hilt.

"Yes," I said, in a nasal accent, stuttering. "As a matter of fact I am quite fond of birds, and this is a particularly good area for them. There is a wide

51

mix of upland habitats here, and this, coupled with the presence of a large freshwater pond, makes it an excellent spot."

He looked at me.

"What are you? Some kind of bird watcher?"

"Yes. I am very fond of birds."

"He likes birds, Mike," the officer repeated to his companion, as if Mike had not heard.

"You like birds, uh?" Mike said. "So what do you see in there?"

Actually I hadn't seen anything but a blue jay.

"Well, in the proper season, it is possible to see all manner of species: there are white-throated sparrows in autumn, black-and-white warblers in spring, downy woodpeckers, brown creepers, black-throated green warblers, blue-gray gnatcatchers . . ."

"What?" he interrupted. "What're you telling me, blue-gray gnatcatchers . . ."

I thought I had blown it.

"There's no blue-gray gnatcatchers in there," he said. "What're you saying? They're a southern species . . ."

"He knows his birds," his companion said. "And his wife's a teacher."

"Evidently," I said, with considerably less flamboyance.

"So what did you see today? Anything interesting?" Mike asked in a friendly tone.

"No, just a few blue jays, and of course the gulls. There's a handsome black-back out there near the island, and there are a couple of cormorants in with the herring gulls."

I handed Mike the binoculars, and he stared out across the water.

"Nice," he said. "Take a look at this, Pete."

Much adjustment, his legs spread, Pete hoisted the binoculars to his eyes and looked at the gulls, focusing and fiddling as he did so.

"See that one with the black wings? It's the black-back. The littler ones with the black wing tips are herring gulls."

Pete was silent for a while.

"Pretty good," he said. "So what's this pure white one?"

"There's no white ones," Mike said.

"There's a big white one here."

"No big white one, Pete. They got the gray. Black wing tips. Those are the herring gulls."

"What d'you call that, then?" he said, handing over the glasses and indicating where he had seen the big white one.

"Jesus Christ," Mike said, just as the radio in their car began to crackle.

"You got a glaucous gull here. Get that call, will you, Pete. Or something. Iceland maybe."

"Let me see," I pleaded.

"This is something. You don't see these much here, I mean it, this is something. I got to tell the wife to get out here."

He was so excited he refused to relinquish my binoculars.

"No kidding, take a look at this. I think this is an Iceland gull."

I would have liked to look, but Mike was hoarding the only pair of binoculars, and I didn't dare point this out for fear of arousing his anger and reminding him of my trespasses. But Pete saved the day.

"We got a ten forty out on 495, Mike, we got to go."

As they were leaving, Pete rolled down the car window.

"Don't walk in there no more. Private property," he said as they drove off.

I thought I was done with my walking for the day in any case, but for some reason, my car wouldn't start. I debated a while and then decided to walk home along the Great Road. I left a note for Officer Pete, in case he came back to see whether I had indeed left.

To get to the main road, I cut across lots, as Thoreau used to say, along the eastern boundary of the Nashobah tract. The land here is half developed with little intrusive dead-end roads with newer houses strung along them. Beyond them is a former dairy farm, which, reluctantly, upon retirement, the owner sold, and moved to Maine to enjoy the remainder of his days, which were not many. He died there a few months later. The farm was sold as a stable, an enormous horse barn was constructed, and many of the former corn fields were fenced into paddocks.

This part of the Great Road, which has been granted scenic road status by the state of Massachusetts (in spite of Mr. Couper's place, some might say), is characterized by nineteenth-century farms, with their barns and farm stands. Not one of the stands is a perfectly maintained boutique-like gourmet shop. There are many homemade signs, some with misspellings, and most with stacks of firewood, old bushel baskets, and drying flower trays and an assortment of farm machines cluttering the pull-offs. About a mile up the road I came to Couper's.

There is something very pleasing about the layout of Couper's farm, the clutter around the front of the houses and barns notwithstanding. The old barn with its sagging roof and its litter of nursing sheds and greenhouses, all weathered out and brown, made an excellent complement to the sweep of lion brown fields, the straggling, blasted apple trees, and the gray line of woods on the ridge beyond.

Mr. Couper was puttering around among his apple baskets in front of the barn when I went by. A whole bushel had tipped over, and the old man, barefoot as usual even in the chill November air, was stooped over replacing them. I knelt down to help him while we talked.

He told me the story of his discovery of the transgressions of Mr. Morrison again, and then, at my prompting, told me what he knew about the day back in 1968 when Morrison shot a teenage girl on his property. I had heard this story before; it's one of the first things you'll hear about if you start asking the old-timers in the town about Morrison.

The event took place on Memorial Day, shortly after the town parade. Two girls who had watched the parade were cutting through Morrison's land to get home, and stopped to rest in the long grass by his garden. The grass was so high and the girls so small, they were hidden from view. Morrison came out on his verandah and saw motion in the grass. He had been having trouble with woodchucks that year, so he went back in the house, got his twenty-two, and without knowing exactly what he was shooting at, let fly.

One of the girls told me later she heard the noise and thought firecrackers had gone off. Then she realized she couldn't feel anything in her legs. The other girl understood what had happened and began waving her hand above the grass and screaming. And then on the heels of this, a mighty storm broke over her head. Morrison, discovering his transgression, was standing over them, cursing them broadly for trespassing. Then he turned on his heels and went to the house and called the police.

"It was a close one," Mr. Couper said. "An inch or two to the right or left, I forget which, and she would have been paralyzed. She recovered though. Had babies, but she had to spend three days in the hospital."

He paused and looked off into the wide blue yonder. I sensed what was coming.

"I been in the hospital myself a couple of times," he said dreamily. "Had a little plumbing problem . . ."

I tried to cut him off.

"Didn't you write about that in your memoirs?"

"I did indeed . . ."

"I read that, quite an ordeal, that was a very good story, and I usually hate hospitals, I'm terrified of them actually, but that was a good story . . ."

I was talking fast to hold him at bay with more details. In fact it was not a bad story at all—it was more about a roommate, a nonagenarian, who, although conscious, didn't speak for as long as Mr. Couper was in the room. One night when the nurses were changing his sheets with him in the bed, as they often did, after weeks of silence, he shouted out, "What the hell you think you're doing," and lapsed into his nonverbal state once more.

Such is the stuff of memories from Mr. Couper's pit of snakes.

The girl who was shot, now a grown woman, told me she saw Morrison at the funeral of another well-loved old man in the town. She approached Morrison, identified herself, and asked him if he remembered her.

"I'll never forget you," he said.

Mr. John Eliot, the founder of Nashobah, was born in 1604 and came to the colonies in 1631. Fifteen years later he began preaching to the Massachusett Indians in their own language, having learned Algonquian from one of the natives around Natick, where he set up his first missionary church, the center from which, as he phrased it, the cords of Christ's tent spread over the land. By 1658 he had completed the great work of his life, the translation of the Bible into Algonquian, and it was published in New England in 1661, one of the first books printed in North America.

Nashope was the fourth of Eliot's praying villages. It was a place of much "affliction," he wrote to the commissioners of the Society for the Propagation of the Gospel Amongst the Poor Blind Indians in New-England, the group in London that supported his work. It was the chief residence of the sachem of the blood, Tahattawan, who had become a faithful and zealous follower of Christ. Tahattawan died shortly after the founding of the village and was succeeded by his son, John Tahattawan, who proved "vain," according to Eliot, although eventually he became an expert in the Scripture and was able to read passages from Eliot's Bible. It may be that he had simply memorized them and was pretending to read. Preliterate people such as the Algonquian tribes had a remarkable ability of memorization and oral skills. John Tahattawan himself died a short while later, and by 1670, when Eliot sent a pamphlet describing his work to the society in London, Nashobah had no ruler. The teacher was John Thomas Good Man, whose father was killed by raiding Mohawks in 1654 while he was fishing at his eel weir on Beaver Brook, the main stream in this region. The site of the murder was close to the present-day Great Road, which was a main Indian trail and was haunted by Mohawks during their periodic raids to the area.

As a result of the violence, the Christian Indians packed up and deserted Nashobah for a whole year—which among other things says something about the nature of the village there. Weaker tribes of Indians tended to run away from problems rather than face them, so that if there was trouble with the Mohawks in an area or, for that matter, conflict with a leader, the bands would simply collect their goods, burn their wigwams, and move on to another site. One of the hopes of the English in converting them was to get the Indians settled in permanent systems, to lay order on what the English must have seen as an apparently chaotic organization.

The fact that the people deserted Nashobah suggests that, in the beginning at least, the structures they had there were traditional wigwams—bark and skins laid over a rounded framework of saplings bent over and stuck in the ground and easily replaced. The structures fitted the weather. They were cool in summer and warm in winter, whereas the heavy, permanent English structures were cold in winter, hard to heat, and hot in summer. More to the point, the flea and louse populations in the English homes could get out of hand. But whenever there was an infestation of pests in their traditional structures, the Indians would simply burn the house and build another. The women did the construction, and they could put up a house in a day.

The meeting of a mobile hunting-and-gathering culture that practiced swidden agriculture and a culture that tended to fix itself in one place for centuries was the source of confusion and eventual conflict. The understanding of land, of the universe even, of the English and the Indians was almost diametrically opposed. Within twenty-five years game began disappearing from those regions where the English dominated. Within thirty-five years, there were serious squabbles between the two cultures, many of them over land, and within fifty years, there was an outright war. Perhaps it was inevitable.

Part of the conflict was the law. This system of inscribing and documenting an abstraction came to the New World with the English as part of an ancient Judeo-Christian tradition that held the word, and most especially the written word, was veritable god and law. If it is written, it is true. And so they inscribed—endlessly. If necessary, they cited sacred texts. Cotton Mather, for example, in his various literary rants against sins such as dancing, cites the Bible to make his points, much as a modern-day born-again Christian would do or a Muslim zealot with the Koran. No one would ever, by the way, accuse the English of not knowing how to use their language, and the Puritans were among the most literate people on earth. Literacy rates in the Bay Colony were higher than they were back in England.

By the 1670s this Puritan concept of written law, of a higher doctrine, had become so established that during King Philip's War, when the wife of one of the sometime residents at Nashobah was killed by a passing Englishman at Hurtleberry Hill, the town fathers, finding the white man guilty under the aegis of town laws, felt compelled to hang him.

This is not to say that the native people of the Americas did not also have a concept of law or, for that matter, a concept of division of land. Territory was defined, and periodically redefined. Generally, the boundary would have been identified by a natural topographic feature such as a watershed or, in the case of Nashobah, the land between two ponds. The territory would have been under the somewhat loose control of a powerful figure, a "king" as the English phrased it. Among the Eastern Woodland people, the social structure was a complex hierarchy not too far removed from the protofeudal system that existed in England before the coming of William. At the head of the group was the sachem and his wife, or wives. This man, the equivalent of the lord or earl in English culture, was in control of a certain territory, a tract of land defined by natural boundaries and comprehended by all those tribes and bands in the general area. Periodically, at a great council, the sachem would divide up his territory and assign certain areas to certain families for hunting, or fishing, or agricultural use. No one owned any of this, though, not even the sachem—he or she merely controlled the rights of use, the *usufruct* of the region. In return the sachem was given a tribute each year by the people below him, a certain number of bushels of corn, for example.

Beneath the sachem was a sagamore, a sort of nobleman or lady, who had rights at the council and who acted as a chief advisor in matters of war or hunting. Below the sagamores were the respected families, the nobility, and below them, the band of families that made up the tribe. Some tribes even had serfs or slaves.

Within this territory, or "kingdom," small bands, extended family groups, or tribes had rights of use of planting fields or hunting grounds, fishing weirs, or berry-picking areas. But they did not in any sense own the land in these areas, and after some years they would abandon "their" fields anyway and move on to another area. All this was ill-defined, so that any individual who wanted to collect sedges near someone else's fishing weir could do so.

Anyone who wanted to dig groundnuts or collect bark near someone else's berry-picking grounds could proceed. Furthermore, at certain times of the year, in certain places, the controls were relaxed and people from various tribes would gather with other bands, usually around good fishing sites. For example, in Sarah's time, the falls of the Merrimack at what is now Lowell were under the control of the great sachem Passaconway (who, it is said, lived to be 107 years old and whose father was a bear). During the spring runs of migratory fish, villages from all over the region would gather at the site to share the bounty. They all acknowledged a mutual right to use the site for a specific purpose even though the falls were in the territory of Passaconway. Under the Indian system, property rights shifted with use.

The law was flexible, and at once personal and communal and imbued with many variations. A good speaker, someone who could hold forth, could get his way. Indians worked by council and consensus, and one of the characteristics that defined their leaders, other than bravery in war or lineage or hunting skill, was skillful oratory. It was at the council that decisions were made, and until the group assented, the argument would go on, sometimes all night, which is one of the reasons contemporary Native Americans in the region question whether Sarah Doublet, a lone individual, could have turned Nashobah over to the English.

By the seventeenth century the English were beginning to believe that land could actually be owned as one would own a thing, although even in the freedom of the New World to which they had retreated, there was still a strong concept of common land and public use of land. A purchase of Indian land, for example, did not necessarily mean that the Indians could not hunt or fish on that land even though it was now "owned" by the English. Conflicts over hunting and fishing rights, over trespass and the like, came later in history, after the English established farms. At Nashobah several laws were laid down to protect private property. Indians could not use, without permission, an Englishman's canoe. They were required to knock before entering a house, and of course, they were strictly forbidden to steal—all of which suggests that there was a lot of stealing and borrowing without asking, and a general lack of regard for boundaries and privacy.

Eliot's original documents granting the lands of Nashobah to the Christian Indians are a mere broad description of the place. But in 1686, after the village was supposedly deserted, Samuel Danforth set out to survey the land. Mathematical surveying as we know it had come into use in the 1620s when Edmund Gunter invented a chain, 66 feet long and divided into

100 links, each 7.92 inches long. Surveyors on the ground would lay marks at regular intervals called stations and at the angles—the points or corners. Danforth would have walked over the Nashobah tract with a team, carrying instruments known as rods, or poles, and Gunter's ringed surveyor's chain. Using these tools, he and his partner would have marked off the rough land from point to point, using wherever possible enduring natural features such as large boulders or bodies of water. But they also used veteran trees, with little regard for natural mortality. The large maples and oaks that are cited as turning points in so many of the old deeds are now gone. One of the turnings marked by a large tree in Danforth's original survey of Nashobah is now a paved fork at the end of Nagog Hill Road, for example.

The actual boundary lines of the Indian lands at Nashobah are much discussed in the historical records, mainly because the various English towns began arguing as to which town laid claim to which section of the original tract after the village broke up in 1675. The bounds continue to be argued over today among the mappers and boundary watchers who have an interest in this part of the world because it is believed to be part of the vast corridor of sacred Indian lands that runs—more or less—from the valley of the Grassy Ground River out to the singular peak of Mount Wachusett. Generally, records agree that the Indian lands consisted of a square of four miles to a side, roughly, beginning at a point near the two ponds and running west northwest for four miles, north for four miles, east, and then south to the original point.

All this, the larger territory of the village, is now developed into two or three towns, depending on whose markers and whose research or whose original deeds you are reading. At various points in history, and still today, Groton, Acton, parts of Ayer, and nearly the whole town of Littleton have laid claims to the original site. But most of the tract, it is now agreed, was in the town of Littleton, which was established in 1714.

For those who take an ecological view of land, however, what is the real meaning of these old scrawled documents and ink-splattered maps? Where is the oak tree marked with an *H*? Where is said wall? And where does it all begin? To know a place, to know the real map of the world, you have to get out onto the land and walk.

Chapter Five

In this village, as well as in other Indian plantations, they have orchards of apples whereof they make cider, which some of them have not the wisdom and grace to use for their comfort, but are prone to abuse unto drunkenness.

—*Daniel Gookin,* Historical Collections of the Indians in Massachusetts, *1674, describing Nashobah*

To Have and to Hold

One day in late December, in the woods above Fort Pond Road, I set out to walk across the hills to Mr. Couper's farm and within a few minutes found myself leaning against a tree, daydreaming. While I was there, I heard, among the resident birds, a sudden uprising of squawks and calls, and on the heels of this a blue jay flew into view. This all happened in an indescribably short period of time, and before the flight even registered, a russet-brown streak of something struck through the barred winter branches and launched itself into the blue jay with such force that both fell struggling to the ground. Here, in a tangle of flapping wings, piteous outcries, and struggles, the hawk, probably a Cooper's hawk, proceeded to pluck beakful after beakful of feathers from the breast of the poor jay. As it plucked, holding the still-flopping jay with its claws, it turned in a circle on the ground, leaving a pond-like bed of feathers around the forest floor.

I remained motionless. The blue jay continued to scream even as its life ebbed away, and then in one last mighty gasp, it struck back at the hawk and somehow managed to escape. To my surprise, the jay rose up and, apparently none the worse for the fight, flew to a nearby low branch and began preening. But the terrible brown streak reappeared and hit again, and once again the two cascaded to the ground in a flurry of beating wings—the white-and-blue flares of blue jay wings mixing with the dull browns and buffs of the Cooper's hawk against the tawny forest floor. Again and again the hooked beak dug as the jay screamed and struggled.

Then the flapping grew weak, the calls more strangled, and then the fight was over.

Until the last few decades, when people who do not make their living from the land began moving onto this tract, everything that has lived here since the land first emerged from beneath the ice has gone through this same conflict. It was a way of life for Sarah and her people, day after day, season after season, year after year: the killing and the struggle and the propitiation. It is no wonder that when they killed the deer in her time, they stroked its nose three times and said the words "rest elder sister." The hunt for them was a coopera-tion, an agreement between the hunter and the hunted and the land that supported both, and they would have wanted to make certain that when the deer entered into the spirit world she would inform the other deer spirits to cooperate. And so their law, the Indian tradition, required propitiation.

For all I know, in the end, after they became Christians, when the hunt was complete and the deer lay sprawled upon the forest floor, Crooked Robin, or John Thomas Good Man, or perhaps Tom Doublet himself, hus-band to Sarah, knelt beside the body of the deer, perhaps stuffed some tobacco into its mouth for the journey to the spirit world and perhaps some green leaves for it to feed upon along the way, the idea being that the deer would travel among the deer people of the other world and spread the news that they should give themselves up to these kindly people to use as food.

And then, stroking its nose three times, these Christian hunters repeated the magical incantations of the sacred texts they had learned from their English teachers: "Our Father which art in Heaven, hallowed be thy name, thy kingdom come, thy will be done. . . . and forgive us our trespasses as we forgive them who trespass against us."

By 1668, twelve years after Nashobah was founded, Charles II had granted new charters to lands to proprietors in what would become North Carolina and Rhode Island, Isaac Newton had constructed his first reflecting tele-scope, and Robert Hooke had just published a paper called a "Discourse on Earthquakes."

On the morning of the 19th of December of that year, Sarah Doublet, who was in her mid-twenties at the time and perhaps married to her first husband, would have gone about the day as she normally would. She started her cooking fire before the sun rose over the eastern ridge beyond Nagog Pond, stirred the English cooking pots containing yesterday's gruel of berries, fat, dried corn and beans, hunks of winter squashes and perhaps a few chunks of venison or bear, and started her day. The exact time of what followed is unrecorded, but at some point on that date, the earth in the place that was Nashobah began to shiver. A rain of rocks sailed out of the blue skies, the waters of Nagog churned and roiled and swelled, and the horned water beast who lived most of his time unseen beneath the surface rose up and spirited his vicious hunting tentacles through the drying air. Thunder broke out of a clear sky, a great roaring was heard by the inhabitants of the village, a deep humming noise issued from the earth, thousand-year-old rock walls cracked and shivered, and the ground became as water.

They fell to their knees and prayed to their new gods, to Jesus Christ and his father god, the Chi Manitou, who created all things, to the maker of all things, visible and invisible, the very god of very god. They broke into song, psalms they had learned from their teacher John Thomas Good Man. They prayed aloud in English: "Our Father which art in Heaven . . ." –in Algonquian: *Spiritu, onk na miffinninuog onkeit wofs wahteanuog, noh Lord Manitoo*–and then perhaps ceased their prayers and reverted and marveled at it all and said to themselves, "Hobamacho has come: he is angry . . ." And they looked out over the roiled waters and saw the writhing tentacles of Ap'cinic, and heard the thundering mountain. And there may have been one among them who came out into the clearing with a staff, probably Crooked Robin, who was a signatory on one of the founding documents but a suspicious man to the English since he may have been the powwow for the Nashobahs, one of their "witches" or devils. And Crooked Robin passed among them repeating incantations, dancing and chanting, and shaking his turtle-shell rattle, and making circles on the earth in front of the wigwams with his staff, and spreading the ashes of the fires. And they joined in, and Crooked Robin chanted furiously now to hold Hobamacho at bay and maybe selected from his satchel a few herbal remedies, and these also he spread in the circle and continued to sing–Christian hymns, psalms, prayers, with Algonquian phrases mixed in. And as he spoke and sang and prayed and wept, the rumblings subsided; the winds of heavens stilled, and a vast arc of blue swirled above them; the chickadees and jays began to call once more, and the red squirrels chattered; and out on the as-yet-unfrozen waters, the great circle of white gulls that had been put to flight by the noise settled again to preen and rest. A few cascades of rock spilled down the face of the outcroppings of granite, a single fall, and then only the softer chanting of Crooked Robin, who, having worked his magic, glanced up from time to time as he prayed to check the condition of the sky. And then he too was silent. And they must have gathered in the middle of the circle of wigwams by the still-smoldering fires–Crooked Robin, John Thomas Good Man and his wife Rebecca, the children, John Speen, Sarah, and the others–and they said in so many words: "What was that?"

In the 1660s there were as many as three Sarahs living around the Christian Indian Village at Nashobah. Given the fact that theirs was an oral culture that depended on the memories of the elders to track the ancestral lines, and the fact that the cultural traditions were purposefully destroyed by the English, it's hard to know exactly who Sarah Doublet was. But she was probably the great-granddaughter of Tahattawan, the blood sachem, as Eliot called him, of the Grassy Ground River people. Her grandfather, John Tahattawan, was perhaps less tractable than the English would have liked. He may not have clipped his hair as the English would have him do, he may have spent time fixing it into elaborate coiffures that were, as Eliot knew, a symbol of his power to his people. In defiance of his Christianity, John Tahattawan may even have shaved his hair back from his forehead and allowed a thick raven black braid intertwined with bright bird feathers to fall behind. He wore shell pendants in his pierced ears, no doubt was tattooed, and may have painted his body with red, black, blue, and white pigments from time to time–also in defiance of Eliot. In winter he probably continued to grease his reddish brown skin with the lard of bears that he himself had killed, eaten, and boiled down for fat. He may have danced and strutted and worn armbands and necklaces of claws. In short, he exhibited a certain primitive vitality and vanity, one of the primary of the seven deadly sins of medieval liturgy.

There is a remote–very remote, it should be said–indication of what Sarah Doublet may have looked like. Her aunt had married Waban, the powerful leader at Eliot's first Christian village at Natick. They had a son whose name was Thomas, who would have been Sarah's first cousin and about the same age. In 1686 an Englishman named John Dunton visited the Christian Indian village at Natick–or what was left of it. Dunton met there the king and queen of the village, and left a description of Thomas, Sarah's first cousin. He was, Dunton wrote, very tall and "well limb'd," had no beard, and had "a sort of Horse Face . . ."

Maybe Sarah herself was tall and well limbed, and maybe she too had a long horse face.

Chuck Roth, who laid out the trail of the town land at Nashobah, has

done more homework on the Doublet lineage than anyone else and has come up with as lucid a description of who was who at Nashobah as anyone. Although research on all this is still emerging, he says that Tahattawan was indeed the head of the entire clan and a more powerful figure in the sale of Concord land than he is given credit for. His family was associated with Waban at Natick, and one of his daughters, Nannasquaw, alias Rebecca, married John Thomas Good Man, the teacher for the town at Nashobah. God favored John Thomas Good Man. He lived to be 110 and died in 1727 near Nashobah, not long before Sarah, who also appears to have been long lived (depending on who she was). She may have survived to 100.

If all this sounds convoluted and, after the space of three hundred and fifty years or so, totally insignificant, it should be borne in mind that it would not have been unimportant to those Indians of Nashobah living at the time. As with many preliterate cultures, lineage counted. No doubt someone living at Nashobah, some older woman, or the 110-year-old John Thomas, or perhaps Crooked Robin, could have reeled off the names of the important people in the tribe far back into their own history. And never mind that in this same lineage, along with well-known sagamores and sachems, you would find a bear or two; the division between human beings and nonhuman animals was not as clearly defined as it is today, and women married bears all the time in their folklore.

What can be known definitively from all this is that there was a powerful woman named Sarah living at Nashobah between 1656 and 1736. After King Philip's war, this Sarah Doublet's name appears on the records of several land transactions. There were other Nashobah Indians around at the time, many of them clamoring for this same piece of earth—the land of Nashobah was much sought after, not only by the local English farmers and tradesmen, but also by Thomas Waban, the son of the powerful sachem of the blood, Waban, who had converted and preached to so many souls at Natick. In spite of the pressure to relinquish the land, the English deemed it necessary to deal with Sarah Doublet. Roth believes the fact that this Sarah, whoever she was, could hold onto her land rights in the face of these two powerful forces suggests that she had some authority of her own.

After the death of her first husband, Sarah married a leader named Oonamog, who was sachem of the people at the Christian Indian village near present-day Marlborough. After Oonamog's death, she was back at Nashobah, and here she fell in with an enigmatic figure named Nepanet, whom the English named Captain Tom or Tom Doublet. It is this last name, Sarah Doublet, that appears on the land transactions after 1676.

~~~~~~~~~~~~~~~~~

The whole question of tribes and their locality, of who belonged where and what they were doing in the seventeenth century, is a subject of intense legal debate in our time, primarily because if you can prove, for example, that you are one-eighth Mashantucket Pequot, then you have rights to the land and to the vast, perverse, financially successful (or ruinous, depending on who is playing) gambling operation in that part of southern Connecticut. Some of the Pequots in the area, formerly a poor group of people who lived in cheap shacks and Quonset huts around Ledyard and Stonington, having proved their lineage, and having successfully won tribal recognition at the federal level and then having won a major court battle to permit casino gambling, have become millionaires.

But, in fact, the whole concept of tribe may be an English idea and the tribes may have been more loosely organized than was previously believed. The new thinking among ethnologists is that the Eastern Woodland Indians, the larger group to which Sarah and her people belonged, were connected primarily by linguistic associations that formed in certain regions—watersheds, for example, or sections of the coastline. The basic organization was the family, an extended group consisting of grandparents, uncles, aunts, cousins, and children. These were grouped into bands of people who traveled together to various hunting grounds or lived together near planting fields. Bands consisted of about twenty-five individuals or so; when they got too large they would split off, or some members of the band would go off to join with another band in the region. These family groups and bands were associated with one another by lineage or clans, a sort of vast organization of cousins who grouped together in certain river valleys. These associations, what later came to be known as tribes, had leaders, the sachems, who could be either men or women and who may or may not have been able to inherit leadership. The whole culture was so disparate and each group so unique that it is difficult to say who did what, where or even when, since things changed dramatically after the English arrived and began recording the life ways of the native heathens. Most of the popular images of the American Indian, even to some extent among Indians themselves, come from the culture of the Western Sioux, the great warrior tribes of the northern plains, who put up such stiff opposition to the whites in the late-nineteenth century and gave so many important names to American history. Headbands, feathered warbonnets, beads, buffaloes, wolves, horses, spears, tepees, and sun dances have nothing to do with Sarah's people. They were a quiet, woods-dwelling folk who lived in abject terror of the aggressive Mohawks to the west, the English to the east, the Pequots of Connecticut, and just about everyone else.

Mr. Couper, whose farm I was headed for the day I witnessed the little life-and-death drama in the Nashobah woods, is a member of the Kimball clan on his mother's side, one of the original Puritan families that moved onto the land around Nashobah after Sarah sold off the tract to the Joneses. He is a well-mannered man, and for someone in his mid-eighties is possessed of a sharp, youthful, even playful mind. It says something about the nature of this area, I think, that of the three or four families that closed in on the former Nashobah lands after 1676, four are still in the area, and one of the descendants of Elnathan Jones recently returned to the town to live, as if the ancient draw of ancestral land was too strong for him to have been comfortable anywhere else.

The Kimball family seems to produce long-lived males. When I began poking around the Nashobah village area, there were no less than three octogenarians among them and one active nonagenarian–Chicken John, who for a few misspent years strayed from his born profession of dairy farming to raise chickens, thus gaining his name. Chicken John, who is ninety-three, is not to be confused with John the Baptist, who also strayed from his roots and, albeit briefly, converted to the Baptist faith, thereby deserting his traditional Puritanical family religion and bringing down upon himself his moniker. The youngest of the group is Junior, a man in his eighties who commonly rides a bicycle twenty to thirty miles a day and then refreshes himself by swimming across the local lake. I used to see him at the town beach, a lank figure in perfect form, who would, without pause, stride down to the water wearing a woman's rubber bathing cap from the 1950s, dive in, and strike out for the wilder shores on the west side of the lake without missing a stroke. Within minutes his white bathing cap would be barely visible. Junior lived with a woman half his age on the west side of the town, near the apple orchards of the town of Harvard.

Mr. Couper is not a man who concerns himself very much with the acquisition of property or with the absolute ownership of those lands and goods which American laws describe as his. For some twenty-five years he has lived with a middle-aged family with grown children in his old family house on his property. A few of the old farmers in this community, which has traditionally specialized in fruit growing and vegetables since the area was first settled, would load themselves in their large American-built automobiles and retreat to Homestead, Florida, in winter to continue their trucking operations. Mr. Couper himself owns some cropland in Florida, which he leases out to local growers. Thirty-five years ago, while he was away, he invited a younger couple to live in his house. When he got back, he decided he liked having them around and they have lived there ever since, free of charge.

To make room for them, Mr. Couper moved out of his bedroom into the cellar, a dank, pitlike hole with a dirt floor and a nineteenth-century rough-cut granite foundation, so loosely constructed that snakes and frogs and salamanders used to appear in his boudoir in spring and fall. It was about this time, as if to entice more reptiles to his chamber, that he began collecting toy snakes, some of them huge stuffed animals that draped along the ledges of the cellar walls, some perfect models, complete with scales and eyes, and some of the wilder tropical Chinese varieties that had the effect of mobility and would squiggle and squirm as if alive when you picked them up. Here amidst his snakes, his collection of old rusted garden tools, machine parts, stacks of magazines, newspapers, and books, and his good dog Tiny (a huge black Newfoundland), Mr. Couper dwelt in contentment. Far from exhibiting any shame in his squalid retreat he was, as the phrase has it, house proud, and would invite interested parties, such as myself, to visit his self-styled Snake Pit. Here in this den-like lair, he kept his pallet, where he and Tiny took their rest—a thin, narrow mattress, o'erspread with a filthy nest of shredded blankets and twisted sheets that had not been changed since the year before.

One sad morning in April, while Mr. Couper was planting apple trees at Morrison's property, the woodstove in the makeshift greenhouse at the farm caught fire and the whole place burned, a fine nineteenth-century Federal-style house that had characterized the hillside in that part of the town since the 1840s. Firefighters saved the barn, but the house and the greenhouse were lost. Within a matter of months, with a lot of help from the community and local builders, and a healthy insurance settlement, Mr. Couper was rebuilding, and within a year the new house was beginning to age well. Unlike other new structures in the town, most of which begin as blights on the landscape and take years to settle in, Mr. Couper's house began to have that old, lived-in look within a matter of months. It is nestled among the existing apple trees, next to the old barn, and perched at an odd angle to the road, among the racks of plants and apple baskets and some of the stuffed animals that were saved from the lost Snake Pit—teddy bears, pandas, snakes, huge dogs of the type one wins at carnivals.

Everyone in town assumed after the old house burned that Mr. Couper would take one of the three upstairs bedrooms in the new house. But instead he moved back into the cellar, which he began to fill with his artifacts. In the time that I knew him, the place was much like his old quarters except that it was drier and had solid walls. But the old snakes were there, and the bits of machinery and the tools and toys. He even had a television.

"Best reception in town," he told me, and offered to show me. He snatched back a cloth cover, revealing the shell of a TV he had salvaged from the dump. Just inside the screen he had placed a primitive painting of a rural landscape, also salvaged from the dump. (Many of the old-timers in this town were great dump pickers. It was a tragedy of a minor sort when the town began recycling formally, renamed the dump the "transfer station," and forbade picking.)

For all his apparent poverty, Mr. Couper appears to be a man endowed with some resources. For years, someone was anonymously donating large sums of money to the town library and education funds. In 1996 it was revealed that it was the barefoot farmer himself. All told, he had given over $200,000. The library had a reception to honor him after this came out, and he appeared on the polished wood floors, amidst the perfumed women, wearing a clean shirt, somewhat rumpled pants, and, to everyone's surprise, a pair of shoes.

Even before this celebration, Mr. Couper had begun his chef d'oeuvre, the great autobiography entitled "Memories from the Snake Pit." The work, which is now reaching volume number eight, has much local lore from the 1920s, intermixed with accounts of more recent vintage, some of which are not necessarily historically enlightening, such as the chapter entitled "My Operations." In spite of these slight digressions, the volumes offer a flavor of life in a rural community in the years between the wars.

Part of the reason Mr. Couper has money (or had money—he seems to have given it all away) is that some fifty years ago, he and a few other farmers joined a local investment club. Couper owed nothing on his house, had no children, and needless to say, did not spend a great deal on his wardrobe, and so even with a modest beginning his money began to expand. He was always a skilled farmer, though—he had attended agricultural college at the

University of Massachusetts, and among other things, he knew a lot about apples. Furthermore, he was, as Chicken John told me once, one of the few men in the town who could actually tolerate Black Jack Morrison. And so, some thirty years ago, when Morrison began laying out new orchards at the old Nashobah site, he hired Couper to do the planting.

When Morrison first came out to the area in 1953, the farm was essentially a dairy operation, run by a man named Richard Barrows. Morrison continued dairying and persuaded two ex-nurses, Edith Jenkins and Fanny Knapp, to tend the cows. He also got a license to raise ducks and dug a pond for them below the yard of the house. About this same time, some more orchard land came up for sale when another local orchardist decided to get out of the business. Morrison was a horse man, a retired country squire who was supposed to be resting on his considerable laurels after a successful career in construction. Among other major projects, it was his company that built the old sewage treatment plant at Deer Island in Boston Harbor, where, in 1675, the Indians of Nashobah had been confined during King Philip's War. One wonders if there is not some sort of cosmic irony at work in the area. Bones of the people who once lived on his Nashobah property were turned up during construction.

Whatever Black Jack did he did with passion. With his main career behind him, plus one marriage, plus his interest in cows and horses waning, he decided, at age seventy, to start another career. He had contacts at the agricultural school at the University of Massachusetts and knew several orchardists in the region who taught at the University of Vermont, and so with characteristic tenacity, he began to research his subject. He read for a year, then researched the tree nurseries to find good rootstock, and then, starting with the original old orchard, hired Couper to refurbish and plant new trees.

On the slopes behind the barn, in a section Morrison referred to as block five, Mr. Couper set out Gravenstein, McIntosh, Red Delicious, Black Twig, and Baldwin. A few years later, he replanted the block with Spartan, Empire, Cortland, and a few peaches and pears. In another section, Couper planted the newly developed Macoun, Red Cortland, and Spartan. Macoun was the vanguard of new varieties at the time and now is very popular with the local farm stands, but Nagog Hill Orchards was one of the first to put it on the market and for a while was the only orchard that would supply it to wholesalers.

To the west, in back of the tenant housing where Dennis and company lived, and toward the fateful pen of Rick Roth's pigs, Morrison laid out an experimental block of numbered roots that had been recommended to him by his associates at the University of Massachusetts. He also put in some Romes, which are generally the last apple to ripen and hold their bright red fruits late into October when other trees are long since bare. These apples lend a sad sort of flavor to the landscape in these parts on chill October days when the cloud rack hangs low over the wild trees to the west.

Under Morrison's too watchful eye, Mr. Couper continued to do the work of planting. Season after season he would arrive to set out the newly shipped trees—Empire, more Rome, and newer, lesser-known varieties such as Royal Gala, Fuji, Mutsu, and Criterion. He also laid out a peach grove on the southeast side of the tract above the pond, where the trees were sheltered by evergreens. They survived for more than twenty years, well beyond the normal life span for peach trees, especially given the severity of winters in this country. By the time Mr. Couper was done planting, Nagog Hill had some forty acres under cultivation and approximately four thousand trees.

By the 1960s, Morrison opened up his pick-your-own operation and retail stand sales. He hired the former cowherd Edith Jenkins and began selling at the crossroads of Nagog Hill and Nashobah roads, using a scale hung by a chain from the branch of a tree. Crop volume increased, and by the 1970s, he strung up a tent and erected display counters at the crossroads. People began flocking to the area, local people at first, then school groups, mothers with children, whole families on the weekend, city-bound people starved for the country life, and leaf viewers who just happened upon the place.

Morrison's wholesale operation began at this time as well. That too began to grow as the quality of Morrison apples became known around the region. Now many of the local farm stands set out counters of his apples which, no doubt, passing urbanites assume to have been supplied by the few paltry trees behind the stands, but which in fact come from the great sweeping hills of Nagog.

The operation proved so successful that Morrison had to construct an insulated cooler under the barn to hold the thousands of bushels of fruit that rolled in from the orchard at peak season in September. And, as with many large operations, Morrison had to rent beehives annually at blossom time to make certain all the various trees were successfully pollinated. He began building bushel boxes at the farm and was one of the first orchardists to use the system known as integrated pest management, in which insect pest populations are carefully monitored before any spraying programs start up.

The method reduces the amount of pesticides used in an operation and is less harmful to the beneficial insects that occur in the orchard.

What began for Morrison as a retirement project got out of hand. Throughout it all Black Jack was a hands-on operator, a micromanager in the current jargon. He was up every morning at dawn, circulating the grounds, pruning and cutting and directing and shouting. He knew what he wanted done and how he wanted it done, and he would make certain that it was done that very way or there would be thunder breaking over the hills and valleys of Nashobah. Neighbors passing along the two roads that cross the orchard have heard the great guns booming and have seen poor lowly peasants quivering at the knees of the manor lord. All but Mr. Couper of course.

"I don't know," said Mr. Couper. "Nobody in this town likes the fellow, very much. But I never did mind him. Except of course for him spreading those rumors about me. He had no business doing that. He knows I'm no sexual maniac, and he had no right to say so."

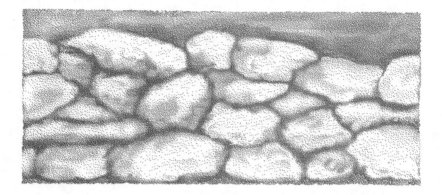

Chapter Six

"Sir Knight, we claim the right to question any man who passes on these roads."

"And from whom hold ye this right?" he asked.

"From one Robin Hood," Scarlock answered. "Of whom ye may have heard."

"...I was in error, it seems, in thinking this forest was the King's domain."

The friar ... stepped boldly into the roadway.

"Dei gratia," he observed, "these woods belong to the King; but de facto, deo volente, they are the realm of Robin Hood. Explained otherwise, Sir Knight, my meaning runneth thus: God gave this forest to his Majesty, and all living things that are found herein he did likewise present to Robin Hood."

—Anonymous, The Adventures of Robin Hood

Common Ground

Early winter in the orchards is a season of stillness and slanting light. The work of summer and autumn is over, the trees are dormant, only a remnant few green leaves hang from the branches like sleeping bats, and the little spikes of frozen orchard grass pierce the powdery snow. The machinery has been cleaned and stored, loppers sharpened and stacked away until the late-winter pruning begins, and Dennis and his crew of Jamaican pickers have long since returned to Clarendon with the proceeds of six months' work hidden away in the soles of shoes and in socks. They lend a strange sadness to the place by their absence.

But early winter is a good time to walk around here. There is a certain clarity in that season that shows up the little stone piles in the woods that are so much argued over by the antiquarians of Nashobah. It doesn't matter whether you're after boundary markers, as some believe these stone piles to be, or spiritual monuments and burial mounds, as others hold, now is the time to find them.

There seems to be a pattern to these anomalous little monuments. Some of the walls, for whatever reason, do indeed line up with the summer and winter solstice sunrise or sunset, although most ramble aimlessly across the landscape, up and down hills and through wetlands. And although I'm never sure I have found them, many of the stone piles do seem to match the turning points on the old seventeenth-century surveys. For my part, I see all these walls and stone piles as frozen time. I see them as potential energy, 77 the stored kinetic energy of some past life, some Indian, or seventeenth-century surveyor or English farmer who, at a given place and time, expended the energy to lift and carry this stone and set it down in this place in that time—for whatever reason. Time is fixed here by these artifacts.

But then something needs to fix the primordial landscape in this area. Development, in the form of huge landscaped trophy houses, is closing in on Nashobah and what we save now is all we'll ever save.

Black Jack Morrison may not be loved by the males in the community, save for a few who actually work with him, but he is much loved by the various women in his life, some of whom he has married, some of whom he has loved without marrying, and some of whom simply work for him. Never, in the years that I have been snooping around the tract, asking questions about the lord of the manor, have I met a woman who would speak ill of the man. The reasons for this, I will not speculate.

Until a few years ago the main contact the public would have had with Nagog Hill Orchards was through Edith Jenkins, who lived on the wild hill above the orchard with her friend Fanny Knapp. These two women, both former nurses, dwelt above Morrison in a sort of reversal of the traditional manorial pattern in which the serfs, thanes, and squires lived on the lands below the chateau or manor house. Morrison reportedly coveted their property, which was a large, undeveloped holding of some ninety-five acres, but he maintained his distance. Both women were great lovers of the outdoors, the type of women who slopped around in baggy jeans and work shirts and knew every bird and insect that crossed their paths. Edith Jenkins, who ran the apple stand, always wore a beret pulled down over her ears, below which stiff, gray shocks of hair protruded. Much to the consternation of Lord Morrison, when the two women grew too old to work, they willed their land over to the Littleton Conservation Trust, which appropriated the property after they died in 1988.

The trust is a nonprofit local group founded in 1972 to save the disappearing tracts of open space in the area. It operates by purchasing land outright, in fee simple, as the Solicitor would say, or accepting gifts of private land which it then opens up to public use. In effect, the trust returns to the town that which in an earlier time would have been the town-owned common pasture or woodlot. The difference is that, in our time, the needs are different. Whereas in the agricultural past communities needed common pasturage and woodlots, today we need–desperately need, in some areas–open space as a counterbalance to rampant development.

At Nashobah this gift of ninety-five acres by Fanny Knapp and Edith Jenkins was the beginning of the restoration of the older order. The land between the two ponds is still generally undeveloped, save for a few inroads that have crept onto the site in the past few decades. Below the former Jenkins-Knapp property along the eastern shores of Fort Pond there are about fifteen narrow strips of private holdings, each containing small summer cottages, most of which are deserted for nine months of the year. At the north end of the pond, still on the former Indian lands, is a summer camp, also deserted nine months of the year, and on the south of Fort Pond there are a few newer suburban houses.

The land around the western banks of Nagog Pond, if anything, is wilder than it was fifty years ago. In the 1920s there were a number of summer cottages along the banks, including an impressive Adirondack-style house on an island just off the Morrison property that belonged to a local doctor. But ever since the 1940s, when the Concord Water Department appropriated Nagog Pond for water supply, the agency has been buying up the surrounding land to protect the watershed. Now, except for the developed eastern shore along the Great Road, Nagog Pond looks very much as it must have in the time when Sarah Doublet lived here–thickly wooded banks reaching down a rocky shoreline.

Flying over the land between these two ponds in the summer when the leaves are out, you would hardly notice any of the modern-day structures or developments. Except for Morrison's orchards and a small clearing on top of the hill where Jenkins and Knapp once held sway and where the Christian Indians may have maintained their garden plots (there is a strange series of regularly placed mounds in the clearing that some local historians believe could have been Indian corn rows), most of the tract is still thickly wooded.

But the center of all this, metaphorically as well as literally, is the estate of John Morrison–two large nineteenth-century houses, a barn, outbuildings, workers' housing, and surrounding it all, the flowering hills of apples. Were they with us now, the seventeenth-century English would recognize the land-use pattern immediately–except that it is on a much smaller scale. The basic layout–a large house, farm buildings, peasant housing, surrounding cultivated land, and outlying forestland–is essentially what they left behind in Europe and England.

By the seventeenth century, just before the advent of the Industrial Revolution, English life centered around the village. The village centered around the church, and the houses were clustered on either side of a central road that led to other villages of similar design. Beyond the cluster of housing lay the agricultural lands and beyond them, in certain areas at least, the wild heath, which by the seventeenth century had been much diminished from earlier times when the dense forests of oak, beech, and ash covered the lands between the villages.

This basic pattern, which varied from county to county, had its antecedents in feudal society and its stepchild, the manor house, which developed in the sixteenth century. The feudal system in England was refined and perfected, if those are the proper terms, with the arrival of William the Conqueror. In its most basic form, a village (from the Old French term *vill*) was no more than a collection of houses, barns, and outbuildings surrounded by planting fields within a surround of pasture, and beyond this the wildwood or forest or wasteland. Under the feudal system, the whole of this was under the management of the lord, who was responsible for the safety of his underlings, who had gathered together under his protection to save themselves from the raiding armies of invaders, such as the Vikings or Normans. Small landholders surrendered whatever rights of ownership they may have had to the control of the lord in order to protect their land, their source of livelihood.

By the time of William, the social system was well established. At the bottom were the serfs, who belonged to other individuals and worked the land. Next up the line were the cottars or cottagers, who were responsible for smallholdings, then the villeins, who farmed as many as fifty acres or more. Above them were the thanes, who drew rents in kind from the villeins and who were in turn responsible to the earls or lords, who were in turn responsible to the king. The names and the structure varied slightly, but such was the basic feudal pattern throughout England and much of Europe.

None of these groups actually owned land. They "held land of" whatever or whoever was above them. This primitive organization, so seemingly remote from our own time, in fact is the groundwork from which so much of our property laws originate. As the Solicitor explained, it was not until the seventeenth century that the practice of a private holding of land came into being, and even then the ground was hedged by common laws and statutes. The idea of land as property, as a thing such as a book or car, would take another hundred years to emerge.

On the one hand, here in the United States, laws relating to landholding seem to be an attempt to move away from feudalism toward an independent state of freeholders of land, responsible to absolutely no one, not even the king himself (i.e., the federal government). But on the other hand, there seems to be a certain primitive desire in this country to return to the old feudal system, with the unique twist that every man considers himself a lord of the manor, if not an outright king, even if his kingdom is a half-acre lot in a suburban development.

In a typical feudal holding, by the time of William, some two to three hundred acres around the *vill* would have been cleared from the native forest of beech and ash. Sixteen to twenty families would live in the village, six cottars or so, maybe nine villeins, and the thane. All told there would have been about two hundred people in the town. The system worked communally. These families would have owned a number of plows among them, possibly as few as seven or eight, and they would have had teams of oxen, also shared, to pull the plows. They may have had community fishponds on the local streams, and weirs, and even a water mill. The field, which began at the forest edge and ran to the edge of the village, was one long, open stretch. The patchwork division of small fields and pastures that you see today flying into London would come later, in the seventeenth and eighteenth centuries, with the acts of enclosure. The great open field was plowed in strips that were roughly ten times as long as they were wide. This pattern, known as a furlong–a standard furrow's length–came to pass because of the difficulty in turning a team of oxen. The long strips of arable land were planted to grain, barley, and peas, and were altered on a three-year system of rotation, allowing some strips to lie fallow in any given year. Each villein planted and harvested his own crop on a given amount of land, but it might not be the same piece of land each year. Under this system, fields of different quality would be equitably distributed among the farmers over a period of time. Unless you were a serf–essentially the equivalent of a slave–you would be guaranteed a certain amount of land and the distribution of these arable lands was decided each year at a meeting known as the annual allotment.

In addition to a share of the field, each family would have maintained, close to its house, a small, privately cultivated plot for a garden, a yard for hens and geese, and a few fruit trees. These were not unlike the small highly fertile, personal garden plots maintained by Russian farmers during the Soviet era.

Surrounding the cultivated lands were the pasturelands, where each day the herds of cattle, sheep, and goats were driven to graze. These lands were also held in common by the village but were not divided into lots. Beyond the pasturelands was the wildwood, which was held, in effect, by no one. Here the local peasants went to gather nuts and firewood, here they turned out their swine to forage, and here also, up until the coming of William, they hunted deer and boar for their larder.

Winding through dark overhanging thickets of beech and ash and holly were narrow little pony trails connecting this system of villages. But one did not normally venture far from the homelots and the outlying pastures, for inside the tangle of this dark forest lurked spirits and wood goblins who would snatch at you from the snagged, moss-hung branches, eat out your heart and liver, and perform unimaginable rituals and ceremonies. The greenwood, essentially the wilderness in the original sense of the word, was the known domain of hideous imaginary creatures as well as all too real escaped criminals and mental deficients who had been turned out of society.

It was also the home of the green man, a mythical figure who was part animal and part human and lived at the forest edge, just beyond the pasture and the plowed lands. He was a landless figure, a primordial character out of the time before agriculture and the strict division of land. He was also benign, a savior of lost children, and the peasants of William's time commonly left out offerings of milk for him at certain times of year to keep him happy. Country people continued to propitiate him up into the early twentieth century. He had many antecedents, including Enkidu, of the Sumerian epic Gilgamesh, and he had many descendants, including the outlander, free-spirited, anti-authoritarian, landless figure known as Robin Hood or Robin of the Wood. Well into the twelfth century and beyond, this Robin Hood and his merry band of men, who dressed, it will be remembered, all in green, maintained the Anglo-Saxon spirit of the days before the Normans invaded and began the long advance toward privatization which, it could be argued, ended with the takings clause of the Fifth Amendment of the U.S. Constitution: "nor shall private property be taken for public use without just compensation...."

William, as Anglophiles to this day will attest, at once altered this primordial feudal system and refined it to his liking. One of his earliest violations of the traditional Anglo-Saxon system was to declare the forest his private hunting domain. Villeins, serfs, and cottars who were discovered in his greenwood collecting faggots, digging out rabbit warrens, or worst of all, killing deer—*his* deer—were severely punished. Their hands were cut off, their ears cropped, and in some instances, they were put to death. William's ruthless protection of his resources altered the ecological makeup of the forest in those areas where it was heavily used by the peasants. It was customary for them to pollard the trees of the woods and to allow swine to uproot native vegetation in their search for nuts and roots, for example. Removing the peasants from the forest may actually have had a beneficial ecological effect, at least around the villages, but it was not good for the people of the region.

(There is an interesting, albeit tragic, contemporary twist to this in the recently privatized forests of Siberia. Formerly, the state would drive out and sometimes even kill individuals attempting to exploit the state-controlled forests. Now it's up for grabs, and one of its most important predatory inhabitants, the magnificent Siberian tiger, is on a swift path to extinction.)

The next thing William did was to assess his holdings. He sent out his scribes, who laboriously, plot by plot, documented all that was owned by William and set it down in the great Domesday Book. Ten years later, King William died, and the ownership of England passed to his son William, William Rufus. Under William the Conqueror's feudalistic system, rents for lands were paid in kind; that is, you supplied a certain amount of grain to the earl each year according to the amount of land you were using. You rendered unto the lord a certain amount of work each year, depending on your landholdings. You applied each year to renew your holding and the terms of your arrangement were set. Rights of use of land formed a great theoretical pyramid, with the king at the top, the serfs or cottars at the bottom, and various tenants and lords in the middle reaches–from the Crown, all titles flow, as the phrase has it. The system was not just designed to control land of England. It was also a convenient way of raising an army. The lords owed allegiance to the king, and the villagers could pay their rents by doing military service. When the king called to raise an army, you joined. So did your lord.

All this more or less came to an end about the time that the Puritans came to the New World, and the old tenure system requiring payment in kind or in personal services faded. The king granted the lands of the Massachusetts Bay Company in *common socage,* which is to say the right of use of the land could be paid in rents, rather than knights' service to the king. Common socage was actually not an unusual form of payment for land in Kent and also in East Anglia, where many of the Puritans came from and where the feudal system had less of a footing than in other sections of England.

Even before 1620, peasants in England were able to maintain certain rights under what was known as the allodial system, which had been in practice as far back as the Roman period elsewhere in Europe. This held that no matter who was in control, no matter what king sat on the throne or who was lord, the peasants would continue on their traditional lands. There were no laws stating this, it was simply a reality, but it was such an enduring one that it has been at the root of the private property system even into our time. With the advent of feudalism, in much of Europe the allodially held lands were placed under the protection of a powerful lord. But in England, and most especially in Kent, from whence many of the New England settlers came, the allodial system was maintained even after William's time. As a result, when the seventeenth-century Puritans began taking over the Indian lands of New England, they understood, perhaps better than any other invading culture of the Americas, the rights of Indian title to those lands that the Indians were cultivating.

Even without its civil wars, regicides, interregnums, and the Puritan exodus to the New World, the seventeenth century was an active period in English history. Not long after the Great Migration to the Americas began, Parliament passed a statute switching all existing tenures into common socage. Then, at the beginning of December 1664, while Sarah Doublet was living at Nashobah, two men at the upper end of Drury Lane in London were reported dead and two physicians and a surgeon were sent in to determine the cause. There had been rumors abroad that the dreaded plague had returned to Holland, and the authorities wanted to make certain that it did not reach England. The physicians inspected the dead bodies and found upon them "tokens" of the sickness. The case was dutifully reported to the parish clerk and the weekly bill of mortality printed the news the following week.

Over the next month cases began to appear with more regularity in the different parishes: twelve in St. Giles by the tenth, twelve more by the twenty-fourth. Seventeen cases in Saint Andrews between the third and the tenth of January, twenty-three more by the end of the month. Slowly, parish by parish, the number of cases mounted until by the end of June 1665, as many as a thousand people a month were dying in some sections. All that summer the infection raged through the city, and by September it was killing seven to ten thousand people every week, some thirty-eight thousand in the two-month period between August 1665 and the end of

September.

Finally, to cap things off, on September 2, 1666, a fire began in the city and spread quickly among the small shops, churches, and livery halls. By the end of the day it was still spreading, and by the morning of the second day it had jumped deeper into the city. For four straight days and nights, the conflagration raged through the maze of streets and shops, and when it ended, finally, it had destroyed an estimated 13,200 houses, some 87 churches, and 44 livery halls. The whole city lay in ruins, commercial centers and administrative buildings smoldering. But within a few days the plague numbers diminished, and by the end of that autumn the disease had died out.

The ashes of the Great Fire had hardly cooled before two architects, Christopher Wren and John Evelyn, appeared before Charles II with plans to rebuild London on a grand scale, based on contemporary town-planning principles. London at the time was a warren of narrow streets and crowded wooden structures. Two of the major thoroughfares, Thames Street and Threadneedle Street, were only eleven feet wide. The lanes, alleyways, and walks were lined with a multitude of houses that crowded one upon the other in such profusion that the inhabitants rarely saw the sun and lived ever in an "unwholesome" shadow. In order to prevent this dangerous development pattern from recurring, after the Great Fire a series of acts were passed that established four house types and sizes, all to be built of brick and laid down with minimum safety requirements about party walls and overhanging jetties.

Anyone living in the American wilds of Utah who proclaims grandly that his land is his to do with as he will still has to contend with the end result of the Great Fire of London. Like it or not, we live on top of the past, under the English system of common law, and these early English codes, organized to protect the safety of the people, were the prototypes of zoning acts and land-use codes and were as much a part of the traditional roots of American land-use law as the Fifth Amendment. One could argue that the Fire Building Acts were a curtailment of the rights of private property (so much as they existed in the England of 1666), and so they may have been. But the end result of the meeting of Wren and Evelyn with the king was not only the creation of the London that is so beloved by the international visitors of our time, but also the beginning of zoning, which, as many still argue, was the end of freedom.

Some years ago, during a little walk I was taking through the Yorkshire Dales, I spent some time at a small farm in the town of Keld, nestled in the Swaledale. This seemed to me one of the more pleasant sites on the planet Earth, at least in the season that I was there. The high moors swept down in great rolling waves to the little valley of the Swale, and every evening in the fading light, the cry of curlews spirited above the high hills and mixed with the whistles of shepherds moving their flocks down the valleys with the help of their speedy little Border collies. I spent most of my days simply walking. I would take off in the morning, hike down the river valley to a pub I knew, have lunch, and then walk back to the farm in the evening as the shadows crossed the valley.

The farm itself was a small eighteenth-century stone building where the old cowherd and his family lived in quiet seclusion. I used to eat with them in the kitchen—not great food, mind you—white bread and tea, jam and over-cooked meats, but pleasant company withal. I liked very much the son, a crooked-toothed young man of about seventeen with watery blue eyes and a sad smile who seemed to have an appreciation for the myth of America. "Are there many bears in America?" he asked one evening.

"Yes," I said without giving it much thought, "many. They emerge from the forest where I live to tip over people's beehives."

This was not entirely untrue, and I didn't want to destroy his image of the country as a wild frontier.

This same boy gave me my first insight into the gulf that lies between American property law and English law. His family had been living in this same farm for nearly four hundred years, which seemed to me, coming as I did from a country that was not even that old, a very long time to be situated in one place.

"So I guess you'll inherit this farm," I said, "and carry on the old tradition for another four hundred, God willing."

"Oh no, not God. Himself, the lord." He yanked his head southward.

"The Lord God?"

"Na, I mean himself, the lord. Himself what lives in London and owns most of this valley."

Three of the five or six farms in that section of the Swaledale, it turned out, were under the control of a certain family who now lived in London and who leased out the lands to the farmers, and had been leasing out the lands to these same farm families for approximately four hundred years. The leases had to be renewed periodically, and there was no legal guarantee that the lord, the current lord that is, would not decide to renew the lease or sell the property, lease and all, to another party.

"This is a very tenuous situation you are in here. You might not get this place?"

"Well," he said. "Been going on now four hundred years. I suppose it'll go on another four hundred or so."

Chapter Seven

I see young men, my townsmen, whose misfortune it is to have inherited farms, houses, barns, cattle, and farming tools. . . . Better if they had been born in an open pasture and suckled by a wolf, that they might have seen with clearer eyes what field they were called to labor in.

–Henry Thoreau, Walden, *1854*

Cross-Lot Walking

One fine day in May when the bluebirds were warbling from the apple trees, and the sky was sea blue, and the orchard was all abloom with white, sweet-scented flowers, I paid a visit to Rick Roth's pig.

I walked up from the pond through the falling petals, keeping a watchful eye out for Raging Bull, leapt over the wall on the western slope of the orchard, and threaded through the brush to the pig yard. Sir Pig was penned that day. He was lying in a mudhole asleep, but when he heard me at the fence, he gave a mighty grunt and lumbered up on all fours and stood eyeing me, snorting comfortably.

I was in an expansive mood that day, having read the night before Henry Thoreau's descriptions of his forays through the privately held farm lots of Concord. Thoreau was one of the few Americans who recognized the prison of private property. Many writers on the subject of land, from John Locke to Thomas Jefferson to Wendell Berry, hold that, in effect, the world is saved by enlightened smallholders who take the business of stewardship seriously and care for the land in such a way as to preserve it for future generations—which might be a good idea if there were some guarantee of good stewardship as well as public access to those desirable but privately held pieces of earth such as the sand beaches and if, as Locke originally proposed, excessive profits from private property were shared. But the world being what it is, there is no way to assure that landowners will not raid their own properties for resources, destroy them for personal profit, and then move on and the devil take the hindmost.

Thoreau did not accept the model of the enlightened smallholder, partly no doubt because he had seen first hand the wornout farmlands of Concord and had visited the timber operations of private entrepreneurs in northern Maine. His singular image of property–at least the one that sticks most clearly in my mind–is his imaginary description of a poor Concord farmer making his way down the road of life, pushing his barn before him. Better to have been born in a pasture and suckled by a wolf, he wrote, than to have inherited farms, houses, barns, and woodlands which you have to spend the rest of your life caring for. By contrast, Thoreau, who certainly loved the land no less than any man, woman, or child, says he has owned every farm and woodlot in Concord at one point or another in his imagination. And as a sojourner on these properties–technically a trespasser–he had come to know them more intimately than the legal owners.

Since this habit of cross-lot walking over people's privately held farms had been my custom around the Nashobah lands for some ten years or so, I appreciated the vindication by no less an authority than Henry Thoreau, and had set out that day with a new appreciation for the land, a sense of who cares who owns what–the world is open for exploration.

Probably it was just the weather–one of the first really warm days of that spring.

The free-ranging days of Homer the pig had come to an end recently inasmuch as, at Mr. Morrison's command, Rick had been more careful about penning him. Staring at his great heaving flanks, I wondered if he missed the old days when he too would set out across the open land to find his dinner on his own, rather than wait in a muddied pen for a slop pail to be dumped into his yard each day.

In the early colonial days the layout of the English farm lots around here would have allowed Old Homer to have his way. Generally, among the so-called outlivers, i.e., those settlers who built their farms outside the town centers, the house, with a small, fenced garden plot and fruit trees close at hand, was the center of the holding. From here, as in England, the land spread out in ever-decreasing cultivation to the woodland, the plowland nearest the house and barns, then the pastureland, and then the wild forest. Pigs such as Homer were turned out into the forest to make their own way; they lived on roots and berries and the mast in autumn. In early winter, they were rounded up for the slaughter. Agricultural historians point out that these swine were hardly great fat-bellied things such as Homer. They were lean, half-wild razorbacks who did not supply that much meat.

I said farewell to Homer and set out for the serpent mound, crossing the wall and keeping Morrison's house on my left. No one around today. Dennis and company seemed to be working in the orchards over the hill to the north, and there was no sign of the Great Man anywhere.

On my way back through the fallen petals of the apple trees, at an alcove near the wall, I nearly stumbled on a couple in an amorous embrace. They were most embarrassed and began explaining that they did not mean any offense and had no intention of doing anything "outrageous," as the gentleman put it, but it was such a fine day and they had been walking on the road beside the pond, which they presumed (accurately) was public, and had just wandered up into the orchard, and it was so pleasant a place and indeed the air so "redolent of flowers," as he expressed it, that they had been swept up with one another, and so on and so forth, in so great an outpouring of verbosity that I did not have time to explain that I was not the owner, but a mere sojourner here myself. It turns out they were bicyclers out for a spin and had stopped by one of the walls and wandered back among the trees, where, it appears, they were overwhelmed with love and swept into one another's arms.

As an inveterate wanderer myself, I had no interest in interrupting their tryst and made signs to move on. But the man, a lank fellow with a shock of black hair over his left eye, was a great parleyer and seemed anxious to chat. His companion, a small blond woman with a direct, unembarrassed, in fact slightly offended stare, waited patiently while her gentleman friend jabbered on, scraping and bowing, and throwing his arms akimbo in the Italian style. In time she too warmed, began talking about local animals, and how they had seen a coyote ten minutes before I arrived. They were themselves fellow travelers, I suppose. It was their custom to come out from Cambridge, where they lived, each fine weekend, park their car in some unlikely place, and ride the back roads looking for good sites. These were not the fanatical type of bicycler that you see in these parts, who commonly apparel themselves in tight-fitting spandex and space helmets with blackened visors and ride furiously up and down the hills of the hinterlands. These two were dressed in khaki shorts with wide pockets, and the man was wearing a white, collared shirt, open at the neck. He could have been an attorney for all I knew.

Their idea of an outing, they said, was to pedal along until they came to a likely spot for exploration, and then hide their bicycles in the brush and bushwhack through the woods.

"The only way to find interesting spots," my new acquaintance said. "You've got to get off the beaten byways and highways."

I was anxious to get down to the area around the purported serpent mound to do some snooping around myself, so I made my excuses and, after proper farewells, cut over the hill into the woods and walked down through the oaks and pines, crossed under the crenellated walls of the granite outcroppings, and then climbed up to the top of the head of the serpent and sat down to think.

Not ten minutes into my reveries, I heard trampings and whispered voices and saw this same couple weaving carefully through the brush. God, I thought, they're going to see me and think I'm some kind of voyeuristic sexual maniac who sneaks around the landscape to spy on lovers. My first inclination was to escape and leave these two to themselves, but they made straight for the hillock—they had an uncanny eye for good spots—and I had no choice. "Ah ha, you again," the man said. This time the stare of his woman friend was indeed icy.

I decided to try to break the ice and began telling them what I had been learning about this site. I showed them the serpent mound, and we walked over to the purported stone turtle, and I told them the story of Nashobah and Sarah Doublet and the crimes of this Mosely, and my attempts to walk the boundaries of the original Indian village.

All this generated a curious coincidence. The woman asked me if I had ever read a book about an area called Scratch Flat, which lies about three miles northwest of the Nashobah village. I knew this book very well, I explained, having written it myself.

Thus transformed from potential pervert to Well-Known Author, I was immediately invited by the couple, who introduced themselves as Timothy and Nancy, to join them later for a drink. We spent the rest of the afternoon walking around the tract and then repaired to Concord to lift a glass.

In the course of the conversation it came out that Nashobah was one of their favorite trespassing sites. They had found it some years back, but did not like to use maps for their explorations and would be unable to find it again, until they happened upon it once more by accident.

"You don't like maps?" I asked. "Explorers who hate maps?"

"They're so boring," Miss Nancy said. "So predictable. You go off with a map, you know where you are, there are no surprises. No adventures."

"No good places," Timothy said.

"Yes, but if you ever do find a good place, how do you get back?"

"Chance," they said. "Luck. Exploration."

Since they had been in the area on their bicycles recently, I asked them if they had ever seen a big blue-eyed man with a shaven head in the orchards.

They thought for a minute, then recalled a certain bull-like individual who shouted at them to get out of his woods.

"He was most rude," Timothy said. "He didn't know we love peace."

"I remember him," Nancy said. "Wasn't he a nice guy? Wasn't he the one who gave us the apples?"

Timothy tipped his hand from side to side and blew out his lips. It turns out that this Morrison, as was his custom, had come storming up from his apple trees at the edge of the woods and told my new friends in so many words to get lost.

"He was not very good with the English language," Timothy said. "He found it necessary to employ many, how shall I say, *expletives,* in order to inform us that it was his private property. But then, having barked, he began to wag his tail, explained that he had a problem with people just coming in and stealing apples and breaking branches and the like, and so he had to start kicking people out. Everyone he saw. Except the neighbors. 'Neighbors OK,'" Timothy imitated. "'Me know them. Good. Don't know. Bad. Kill. Eat.'"

"He wasn't anything like that," Nancy said. "He was a nice guy, gave us apples. Just an old duffer with a big bark. You've got to get to know these guys."

"He was not what you might call a poetic man," Timothy said.

I couldn't help wondering if part of the reason Raging Bull was behaving properly—more or less—was that this Nancy was very pretty, in a traditional sort of way, blond hair clipped off at the shoulders, and big china blue eyes that could, as I had noticed earlier, wince down to little ice-blue darts. But even Timothy admitted that Morrison was just a large dog defending his territory.

"What can you expect? A place like that. The Greeks often placed monstrous guardians at the very gates of their mytho-poetic sites. He was just doing his job, I suppose. The stalking, glint-eyed swordsman guarding the Golden Bough at Lake Nemi."

Given his metaphor, I had to tell them that I had heard from one of the archaeologists with whom I had walked this tract that some of the best-preserved Native American sites are often owned by individuals who are rabid protectors of their properties and forbid all trespass.

"That's a sacred site, you say?" Nancy asked.

"Probably," I said. "The conditions are right, the presence of water, 95 south-facing slopes. There is a theory that the Christian Indians selected their traditional sacred sites for their praying villages." I explained that the Nagog Hill would thunder periodically in this spot and that the local mystics and antiquarians held that the spirit force was very strong there.

"We should have known," Timothy said, nodding to his companion.

"Mmm," Nancy said, in a very comfortable manner.

"How should you have known?" I asked.

They looked at one another again.

"We are often consumed by love in that place," Timothy said.

I went to a wedding that spring in the nearby town of Shirley. The service was held in the church on the old town common, which is one of the least-spoiled town greens in the area and is on the National Register of Historic Places. For some reason, this small town of Shirley has been home to a number of people very much involved in land-use planning in the United States. Benton MacKay, the man who conceived and laid out the Appalachian Trail, lived there, and Lewis Mumford and the town planner Henry Stein had used the village as a model for a documentary film they made about the rebirth of the use of common land in new developments. It was probably a propitious place to get married.

I met an older man at the reception and, given the venue, got into a discussion of land-use policy and private property. I brought up the subject of Linda Cantillon's newly discovered development plans, and the old man visibly stiffened. It turned out he was a relative of one of the sellers and knew Linda.

"This Linda," he said. "All she wants to do is walk around on those fields and look at flowers. We never did mind, really, saw no harm in it. But she was like Ferdinand the Bull, if you know what I mean. She goes out there sniffing the flowers and then one day she gets stung and goes into a rage. I don't get it. A man's land is a man's land. They've got no other way to get money to pay these taxes. They'd love to hang on to it as much as anyone, but they've got to sell."

Linda, who up until this point had concerned herself primarily with the search for new patches of fringed gentian or cardinal flower, was indeed getting into a rage.

She began asking questions at the town hall after she discovered the fate of the land. The developer, a man from the town of Lynn, who had been involved in a number of similar projects and was, according to Linda, a meat eater, a cigar smoker, and the conductor of an oversized gold-colored Lincoln equipped with car phones and fax machines, had offered to buy the land from two older families, the Frosts and the Whitcombs, who had held the land for several generations and now, as the wedding guest had explained, because of the expense of inheritance taxes, were forced to sell. All told the tract consisted of some 113 acres. It was a very attractive property from anyone's point of view, with lakeshore frontage, hayfields, and a mixed woodland, and, save for the monstrous telephone tower constructed on the site some years earlier, as pleasant a place to spend an afternoon as any in the area, save perhaps for Morrison's forbidden fruit orchards.

In this pleasing rural tract the developer planned to construct his trophy houses with two-acre lots, a swimming beach at Long Lake, and—if the town would permit—a privately run mini-sewage-treatment-plant to avoid the nasty problem of poor drainage. One parcel of the land in question had been in the family since the early eighteenth century, and the other was part of the original tract of land purchased of the Indians of Nashobah in the seventeenth century, although it was probably just beyond the boundaries of Sarah Doublet's five-hundred-acre tract.

Why, one might ask, after some three hundred years of successful yeomanship and landholding and economic benefit, should it be necessary for the two families to sell? Therein lies the snag in the American tax system, which, in its wisdom, sees fit to tax landholders for what the assessors *think* the land might be worth based on local market values; that is to say, if some phantom buyer, under the right conditions, were to purchase the land, develop it, and make a healthy profit. As long as the whole community was agricultural, the land was assessed as a farm. But as soon as another, more lucrative use appeared, the assessed value of the property went up. This of course makes perfect sense in a nation in which land is viewed as a commodity that can be bought and sold like a table or a boat, and especially in a region that believes that in order to demonstrate success in life one must have and hold property.

In this country, depending on what state you live in, there are now various means of reducing the taxes on land by placing conservation easements or restrictions on the property, using the land for agriculture, or donating the property to conservation organizations. There are even some state programs that permit landholders to sell or transfer the rights to develop a tract to another section of the community, thereby keeping the land open without financial sacrifice.

These programs do not always allow the landholder to maximize profits on a piece of property, but there are not a few enlightened families around who are more interested in keeping their ancestral lands intact than making money. I know several families in the Nashobah area who cannot bring themselves to sell the old farms even though they stand to make healthy profits.

Furthermore, conservation restrictions work well for buyers seeking peaceful environments rather than lucrative investments. The state of Maine has a conservation program, for example, in which an agency buys wild land and resells it with restrictions on development. One friend of mine of decidedly modest means was able to afford highly desirable coastal property specifically because there were so many restrictions on what he could or could not do with the land. "I actually like the restrictions," he told me. "It means the land around me will always be protected. What do I care if I can't sell it off for development sometime in the future. That's not the reason I want to live here."

This current American attitude toward land as a potential investment is more or less the opposite of that of peasants in the developing vacation areas of southeastern Spain. In Andalusia, smallholders often value a property according to the quality of the crops their ancestors have nurtured over the generations. The asking price for a given plot of land would be based on the size of the grapes or the almonds that grow there, not the potential as a place for a building. The fact that the farm might lie at the end of a rocky, nearly impassable track, with no water and no electric lines and few mod cons, as the Brits say, makes no difference at all. This attitude has confounded any number of retirees seeking cheap land in southern Spain.

⁂

Having determined by a visit to the town hall the nature of the development, Linda composed a passionate letter to the local paper. She was not aware at this point in her career that just because a development has been proposed does not mean it cannot be stopped. And so, assuming a loss, she wrote a sad farewell to what she once thought of as her place. The letter was a paean to wild nature.

She described the tract, which was probably unknown to most of the people in the community, and then, as she explained, she felt compelled to speak out for the current residents of the area, since no one else would. She meant the foxes, the deer, the snakes, the ducks, the sparrow hawks that hover above the high hill, the meadowlarks and the bobolinks that inhabit the grassy fields, the salamanders, the insects, and the frogs and toads. "Where are they supposed to go," she asked, "when their homes are gone forever?" She said she was in mourning for mullein, for the cow parsnip and the turtlehead, and the little woodland patches of bloodroot. Her only hope was that when it was all over, after the native plants of the tract were stripped and lay beneath asphalt, and the suburban landscape of chemically fertilized lawns and thick layers of bark mulch overspread her precious wild gardens, some lone bloodroot would survive and in time spread and somehow heal this wounded place.

"The earth does not belong to us," she wrote at the end of her letter, citing Chief Seattle, "we belong to the earth."

It was a funeral oration, an elegy, an ode to wild nature. But it was delivered too soon.

As any defender of the right to develop private property will tell you, the company of starry-eyed conservationists is legion. Some of them, furthermore, are well armed, and Linda's letter brought them out from their lairs.

She had a call a few days after the paper came out from a man in a nearby town who explained that it was not necessary to stand by and watch the world disappear. There are things that could be done. He went on to describe the use of conservation restrictions, the outright purchase by the town, using state-aid funds to help pay for the land. Could she get a group together, the man wanted to know, who would support the preservation of the tract?

"You bet," said Linda Cantillon.

Within a few months of her discovery, she had managed to rally a few neighbors. Among this group were two powerful women allies named Sarah Foss and Patty Townley. The group began meeting on Friday nights and soon they had enough people engaged in the project to give themselves a name, the Friends of Open Space.

Linda invited the man who originally called her with advice, a local land protection advocate from the next town named David Koonce. He became a sort of strategist for the group in the beginning, although within a few months Linda and her allies had cut their teeth through experience. Quite quickly she came to believe that there were factions in the town other than the developer who approved of developments of this sort because they believed it was a way of increasing the tax base of the community. Furthermore, she began to suspect that other parties had designs on this piece of real estate, namely the town board of selectmen, which some felt wanted to save the property as town land in order to build municipal structures there, such as a sewage treatment plant or a school. This, according to the selectmen, was not accurate. They simply wanted to move cautiously with this open space proposal and so were not throwing themselves whole-heartedly behind Linda's conservation-minded strategies.

As Mr. David Flagg readily points out, anyone who wants to build so much as a chicken coop in this country must now meet a number of zoning, building, and health codes before construction can begin. Permits must be filed and approved and meetings relating to the work must be held. These meetings are open to the public, and attendees are invited to comment, within bounds, on the business of the day. At one of the first of these gatherings, Linda, who had very little experience speaking in public other than parent-teacher meetings, tentatively raised her hand, was duly recognized, and stood to speak. She said, in so many words, that this whole idea was a disaster. "You've got laws here against this sort of thing. He's going to put a beach there?" she asked. "Give me a break, where's the wetlands permit? And what about these bobolinks that nest in the hayfields, what about the blue-spotted salamanders that might be breeding in the vernal pools there in the woods . . .?"

Before she could make her real point, which, of course, was that this whole thing should be cut off at the pass, the chair, politely, suggested that this was a meeting concerning curb cuts for the development, and that this was not the time nor the place to comment on the overall development, nor the character of the developer, nor the overall detriments of the project, all of which Linda had managed to work in during this opening salvo.

"What do you mean I can't comment on the project? I thought that was the point of the meeting."

"It is," they replied. "But only on curb cuts."

"Curb cuts . . ."

"Yes, curb cuts."

"What about the water snakes and the frogs that are going to be evicted if this beach is put in . . ."

"I am sorry, out of order," said the chair. "You may comment only on curb cuts."

And so it went, night after night of meetings in which the group would attempt to use the permitting process to comment on the project as a whole, until they learned that the real way to control this development was to beat them at their own regulatory game and simply pay obsessive attention to tiny details of each permitting problem and fight them, hedgerow by hedgerow, through the infernal bureaucratic system.

"What else can you do?" Linda said. "It's the way these jerks think. They don't get the big picture."

❧ ❦

Two miles down the road from Linda's land, the co-housing group had finally found a suitable place to put in their development. They too had learned, albeit from the other side of the board, how tedious are these eternal permitting processes.

One of the prime movers of the co-housing development at Nashobah was a man named Alexander Bloch, better known to his friends as Sacha, who works as a chiropractor in Brookline and Concord. He and his wife, June Jones, were then living, as yet unmarried, in a small apartment in the city. June had drifted east from the West Coast, where for a number of years she lived on an ashram working as cook. Sacha was from New York City, where he had been writing film scripts and driving a taxi to make a living. Driving, he said, became a metaphor for him, a sort of city street sutra that kept his mind off troubling matters. But after three years, having failed to sell any scripts because, as he said, his agent was always trying to get him to write less *meaningful* stories, he too drifted into the Boston area. They met at a lecture, courted, and moved in together and then, since things seemed to be working out very well, decided to marry.

The ceremony was held at a small chapel that had been deconsecrated and now served as a center for peace and justice. They were married not by a religious but by the director, a man whose group had established a school for children so handicapped that other institutions would not take them. June wore a sky blue Indian sari and affixed around her neck, and on her ears and fingers, many silver pendants and rings. Sacha wore a loose white khadi cotton shirt. The participants sang together and danced in a circle holding hands. June's father, a lank fellow who worked as a railroad man in Colorado, came east for the ceremony and joined the circle with June's mother, who wore a tight blue dress and plastic glasses with sequins embedded in them. Sacha's family, Russian Jews from New York City, came well attired, in double-breasted suits and bright dresses and many jewels. They too joined the happy circle of younger people, and everyone danced and sang, and there was incense and champagne and sacred herbs, and everyone enjoyed themselves, and after it all, in the green light of the late-summer evening, the guests lingered at the chapel as Sacha and June puttered away in their old broken Datsun, with its many fading feminist bumper stickers, to live happily ever after.

Ah, but how tedious is this American dream. What next? A happy life with a little suburban house with a green lawn and a dog and two children, and every day Sacha would drive to work and come home, and June would have the dinner ready and they would eat and watch television together and then go to bed and then sleep and then get up in the morning and then do it all over again. And their friends in the neighborhood would do the same thing. Every Saturday morning, Sacha would dress in jeans and a flannel shirt and go to the dump, while June vacuumed, and they would spend the whole day running the errands they could not run during the week because both of them were now working so hard to get what they had that they never had any time to do the things they really wanted to do. And then finally, on Saturday night, too fatigued to go out, they would rent a video at the nearby minimall and take it home, and have dinner and loll upon the couch and even fall asleep there, dead to the ecstatic night world in the woods beyond their neighborhood. And on Sundays they would read the paper, and then, perhaps, take a walk and then eat dinner together–perhaps, if there was time–and then start the week all over. Again.

"If there's one thing that repulses me," Sacha said, "It is the American dream. I never could see myself living in a house in the suburbs. I can't think of anything more boring."

Sacha and June knew a lot of people in the area who felt the same way, and so in the late 1980s they began holding a series of meetings among the like-minded to talk about an alternative. They had heard about the co-housing experiments that were taking place in Denmark, and the few that had been constructed in the United States, and so they sent out their members to research the matter, drew up a list of what they wanted to see in a community, and then formed a sort of Mayflower Compact that spelled out, in basic form, the social structure of the place. Then they began looking for land.

They found a good property in Acton not far from the local railroad station and began to negotiate with the developer, and things seemed to be going well until the dreaded permitting process began to rear its tentacles and reveal violations that would occur if this project were built, and everything fell apart. So they started over again.

They heard about a fine piece of land near a pond, also in Acton, and Sacha and the lawyer for the group went one day to meet the owner to talk about the terms. Sacha told me about the man.

"He came at you like a bull," Sacha said. "His head lowered, a big meaty hand. Ice blue eyes. A shaved head. Dressed in a flannel shirt. Rough guy."

He sounded familiar to me.

"What was his name?" I asked.

"Morrison something. John, I think."

It was indeed Black Jack himself who, in addition to the lands around Nashobah, owned tracts elsewhere in the region. Now he was ready to sell some off.

"But you didn't buy?" I asked.

"He wanted a million dollars for something like twenty-five acres," Sacha said. "Too much. Absurd really."

Sacha and his partner said—politely—that they would have to think about it, but that it did indeed seem like a lot of money, and perhaps Mr. Morrison, if he were so inclined, could see his way clear to reducing the price and then, if he were to do so, they would be very interested since it was a fine property and just what they were looking for, and they could, perhaps, talk some more about all this, and they would be happy to explain their ideas to him for the project if he would like. Sacha and the lawyer, who was also a member of the co-housing group, were men who were used to social interchange, to nego-tiation, to innuendo, body language, and shades of meaning.

"That's my price," Morrison said and left the room.

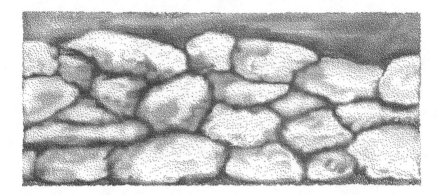

Chapter Eight

Our manifest destiny [is] to overspread the continent allotted by Providence for the free development of our yearly multiplying millions.

–John L. O'Sullivan in The United States Magazine and Democratic Review, *1845*

Terra Nullius

I have a friend who is a fanatical birder, and as such often trespasses on people's private properties. The Concord Water Department lands around Nagog Pond are one of his favorite spring hunting sites. We were talking about the area once when he informed me that there was something "spooky" about the land there. "I don't know what it is, some very bad vibes there."

This is a man who concerns himself primarily with the coloration of the secondary wing feathers of sparrows and finches and is a professed antireligionist. I told him about the events of autumn 1675 in this place. "Figures," he said.

I myself was given a fright here one still night in early spring. I was coming back from a late-night event and on a whim decided to stop off at Sarah's tract, for no particular reason. I parked the car in my usual spot and picked my way up along the serpent mound until I could feel one of the trails below my feet. By this time I knew the tract well enough to get around in the dark, and more or less feeling my way with my feet, I walked north toward the granite outcroppings.

Where the land began to slope northward toward the rocks, I cut off the trail and, threading through the trees, found the rocks and climbed up to a height and sat down. Not that I could see anything. I just thought it would be a good idea to sit and wait.

This was one of those warm, foggy nights you sometimes get in this part of the world, the type of night when the remnant snow patches smoke and steam on the forest floor and mist hangs on the lower branches. In the dim light I could see only the vertical shots of dark trunks and, higher up, the crossed bars of the limbs. The floor of the forest was all swirling with mists and ghostly patches of white snow on the northern slopes. It was still too early for frogs, and except for the occasional scratching or slogging sounds, one of the immense silences that sometimes descend on this place had fallen. Without thinking much about it you could easily slip into the old primal attitudes toward nature that are so much a part of the Anglo-Saxon past and see this as a fenny bog surrounded with snagging, clawed branches and cold, demon-haunted cliffs, where wolf packs den, the lair of boar and bear. You could half believe in Ap'cinic and hear his slicker-slack tentacles searching along the rocky shores of Nagog, seeking human victims. Of course, dark deeds did occur here: the destruction of a whole culture, on the one hand, and on the other the unchronicled rites of the native cultures that inhabited this place over its ten-thousand-year history. Who knows how many Mohawk raids there were in this area before the English began to record them? Who knows how many Pawtucket captives were strapped to this cliff face and slowly tortured to death by the conquering raiding parties?

I was not necessarily meditating on these events while I sat with my back against the dank rock wall. In fact I was thinking about the antithesis of the dark side of the human condition. This region west of Boston is particularly rich in chamber music societies, and I had just been at one of those small, warmly lit events listening to a local string quartet perform. One of the pieces they had played was the haunting Dvorak quartet in F major, nicknamed "The American," composed in the bright summer of 1896 when Dvorak was living in Iowa at the Czech community in the town called Spillville. I was thinking mainly of the mysterious, slow movement, which somehow seems to me to creep across the land slowly, not unlike—it struck me—the serpent mound below me. Suddenly in the midst of this reverie I got very cold, as if a cylinder of winter air had descended. The oddity of the situation struck me. Here I was sitting alone in street clothes in a dank forest, freezing, while not half a mile away the world was putting itself to bed in warm upstairs bedrooms, with clean sheets and well-lit lamps.

On the heels of this the cold became palpable. Something seemed to slam me back against the wall, and I leaped up to resist. It seemed that all the horrid realities of this place, or something more concrete, some definite entity, was there, something so large and terrible that I did not want to confront it. My first thought was to get out.

I moved down the slopes, battering through limbs and stones, believing all the while that it, whatever it was, was still there right behind me. Probably I had been thinking too much about the seventeenth century, when demons and night worms and creatures of the untrammeled fens were not considered mythological entities.

Back at the car I stood on the road for a while looking up at the hill, simply feeling the sensation. Nothing happened, nothing seized me by the throat and dragged me back into the swamps to draw me, struggling, into the murky depths. And below me at the pond, I could not see the slimy gleam of the blind, searching tentacles of the Ap'cinic, feeling along the shores for victims.

I got in the car and switched on the headlights, half expecting to see some darting form cut back into the trees. But there was only the swirling of the mist, the barred limbs, and the dark shapes of the pines out on the island just offshore from Black Jack's orchards.

I had heard from other people who trespass in this place that they had been overtaken with similar experiences—a sensation of cold then an overwhelming urge to run. Local rumor among a few of these people is that Nashobah is haunted. But in all my night walks and misty autumnal day treks, until that night I had never experienced anything there but a certain benign lingering sense of time and historical events, a pleasant richness of landscape.

The original Indian title to this land was deeded over to the Christian Indians by the General Court in Boston on May 14, 1654. Slowly, over the subsequent years, most of this sixteen-square-mile holding was nibbled away at by the English, first by a family named Lawrence from the town of Groton, then by the Whitcombs and the Kimballs and the Proctors, some of whom came directly from England, and some of whom moved up from Concord or eastward from Groton. By 1680 or thereabouts, the original Indian tract was cut down to the last five hundred acres, held by Sarah Doublet. The story is that after King Philip's War, when most of Eliot's Christian Indian villages were disbanded, their inhabitants either dead or collected together in a miserable band at the original town at Natick, Sarah Doublet was told that she could select some portion of the original village at Nashobah as her own. She had perhaps three or four square miles to choose from, but she chose the tract between the two ponds.

There are many theories on why she picked this particular area. The most likely is that it was the site of the original village, but another is that this was originally the sacred land of her people before they became Christianized. Historians of the Native Americans of this period believe that the last tracts of land the Indians would surrender were those that held spiritual significance for them. Hunting grounds, even agricultural lands, may have been sold off or traded, but the last to go would have been those areas that were sacred.

One thing is clear in all this. After 1498, when John Cabot claimed the land for England, the Puritans did not, as is commonly stated, "steal" or "take" the Indian land. They acquired it through legal transfer, either through barter or purchase—never mind, by the way, the parity of the exchange—in other words, trade goods such as beads or knives, or a small amount of money, for a large amount of land. The transactions were legal in the eyes of both the Indians and the English. And therein lies the story of Nashobah. The deeds are there, residing in the archives of the state of Massachusetts, detailed, laborious delineations of what is essentially an undocumentable concept—the wild flow of land and its elaborate weave of plants and animals.

Whether or not the two sides fully understood the meaning of those documents is another matter. The English believed that they had purchased the actual ground and all it contained, in fee simple, the absolute ownership of the place. What was sold, the English believed, was the land itself, an abstract idea whose bounds remained fixed in time and which could be inherited and passed down from generation to generation by the owners. The bounding of this land in this manner meant that in time a different attitude toward that place would develop, one involving long-term care, and shaping, and reconstruction if necessary, in order to make the land fit the needs of its owners. Unlike the native custom, which was to clear an area by cutting and burning, plant it in corn, beans, and squash for a number of years until the soils were depleted, and then move on to new planting fields to start again, the English began clearing land forever. They cut the trees and used them for permanent houses. They twisted out the ancient stumps and lined them up as temporary fences. They cleared the rocks from the new fields, rather than plant between them, and placed the rocks in lines along the fields—to keep out, not in, during this period at least, the free-ranging cattle and swine. They planted with an eye to permanence. They owned this land outright, and since it was owned, it could also be sold.

The Indians believed that what they were selling was not the actual land, but the rights to use the land. This must have introduced a certain irony to the Native American mind, since in many cases, including the transfer of the land around Nashobah, the English at first allowed the Indians to continue their normal habits on the land. They were permitted to hunt and fish, and in some areas they were even allowed to continue to maintain their planting grounds, as long as the English were not using the plot. So from the Indian point of view, the exchange must have seemed more than equitable. In fact one cannot help but wonder that the Indians perhaps believed these English suffered from limited mental capacity. There are records of Indians selling land two and three times over to different parties, since, in their minds, they were not selling the land per se. It was common for sachems to distribute rights of use of a territory to more than one group.

This created an anomaly that eventually led to serious breaks between the two cultures. The English may not have understood that the Indians did not understand, and the Indians did not comprehend that the English did not understand the concept of usufruct. It was all a grand *mis*understanding that, in the course of a few years, would lead to war and in the course of a few centuries would lead to the total destruction of ancient American cultures and the loss of almost all the land of the continental United States, Canada, and Mexico.

All this is written. You can see the record of the land transactions in the scratches and scrawls, the crossings-out and ink blotches and Indian marks, in the deeds and surveys describing the angles, the rocks and stone piles, the old trees, and the hills. Here, in the registry of deeds, in a thousand courthouses across America, you will see the record of how places such as Nashobah were created and how they ceased to be—and how, by extension, America ceased to be and came to be.

The ethics of taking over the lands of North America by the English and the French was based on the belief that North America (and, later, Australia) was in fact *terra nullius*, that is to say, land owned by no one. In New England land was granted outright by the Crown, which derived its claim from Cabot's "discovery," as it was believed, of the region. The Puritans, who for all their faults were really the most ethical of the colonists of the New World, believed that since the Indians were not clearing land to farm, they did not use the land and therefore were not in possession of it. The logic of this was clearly spelled out in the Word, in Genesis 1.28: "Be fruitful and multiply, and replenish the earth, and subdue it . . ." By contrast, if the Indians had cleared land for planting fields, the Puritans felt obliged to pay them. The fact that the Indians depended on use of the wild, uncultivated lands, and that this was their land, their place even if it was a wilderness, did not occur to the Puritan mind.

This unique, somewhat bizarre, custom of one culture not even recognizing the existence of another was turned upside down a few years ago by a Chippewa acquaintance of mine named Adam Fortunate Eagle. On September 23, 1973, he boarded a plane he had christened "Chief Joseph" and took off to claim Italy for the Chippewas. Why not, he reasoned, if the purported Italian navigator Columbus could do the same for the Americas?

The whole gambit was a setup, of course; he had even informed the Italian consulate in San Francisco of his intentions, and to their credit, the Italian government and the ever hungry papparazzi took it very well.

Knowing he might encounter "native inhabitants," perhaps savage, Mr. Fortunate Eagle arrayed himself in ceremonial attire for the big event, a fringed buckskin war shirt, a silver turquoise ring, and a traditional feathered headdress.

At dawn on the 24th, the "Chief Joseph" descended from the clouds and landed upon flat ground beyond a beautiful Italian city that the natives called, in their language, *Roma*. Having got word of his arrival, the locals were excited by the appearance of the great man, reporters surrounded him, fawning, but he strode through the crowd and raising a ceremonial spear, drove it into the Italian soil and thereby claimed the land for the Chippewas.

"Take me to your leader," he said to the assembled.

He was led to a grand palace, where he ascended a wide flight of white marble stairs, and was ushered into a spacious room where a small dark man in a gray suit was sitting. The small man rose and greeted Mr. Adam Fortunate Eagle warmly, clasping his right hand in his and moving it up and down briskly—an apparent greeting ritual of this culture. Fortunate Eagle was informed through interpreters that this man was Giovanni Leone, the current sachem, or president, of the Italian nation. The two chatted amicably, and while they talked a message was delivered by a page. The spiritual leader of that country, a sort of chief medicine man, known as "The Pope," had heard of the arrival of Mr. Fortunate Eagle and wished to pay homage to the great explorer.

At eleven o'clock the following morning, Adam Fortunate Eagle entered into a grand city within a city. It was a vast, jumbled collection of buildings almost, but not quite, equal to the great cities of Teotihuacán or Chichén Itzá or the mountain retreat of Machu Picchu, constructed on the American continent by Mr. Fortunate Eagle's fellow Americans some years before the construction of the Italian edifices. Inside the buildings were windows of colored glass and many statues of people and long hallways of shining polished stone. The proud natives conducted Mr. Fortunate Eagle to a room with an arched ceiling, covered entirely with paintings of human beings—some equipped with wings. In the center of the ceiling was a painting of a scowling man with a white beard, his arms outspread. This, the natives explained, was their chief god, and this room, this chapel, as they called it, was one of their primary cultural centers.

Mr. Fortunate Eagle, still clad in his celebratory dress, was mildly impressed with their temple. But he had also seen the frescoes at Bonampak and the paintings on the walls of his own temples and could not, for all the world, see what all the fuss was about.

Then Mr. Fortunate Eagle was conducted by an entourage of splendidly caparisoned officials into an inner sanctum where their chief shaman held court. A door opened and a man in white vestments and a conical hat came forward and greeted him. Adam Fortunate Eagle was informed that this was the famous Pope.

The Pope held out his hand, indicating by signs that Mr. Fortunate Eagle should kiss a bejeweled ring he wore upon his finger. The discoverer of Italy stared briefly at this impostor, and then held out *his* ring, the silver one, with the precious turquoise stone, indicating through signs that the Pope should kiss it. Instead, the Pope laughed heartily and took his hand and held it while the two of them spoke of their two nations. Speaking in English and with apparent sympathy, the Pope said he understood the plight of the Chippewa people and the American natives.

"Thank you, my son," said the great discoverer.

Adam Fortunate Eagle returned to the Americas and announced his discovery and there was a little flurry—for fifteen minutes—in the American press. The story was even written up in *Time* magazine. But it didn't do much for the plight of the American Indians. That story was written too long ago to change.

England, of course, was not the only country vying with the Indians for control of their continent. Nearly two hundred years before the founding of Nashobah, the Spanish began setting down colonies in the Americas. Spanish law had evolved directly out of Roman law and church law and the land-use patterns differed considerably from the settlement patterns in New England, or for that matter the English settlements at Virginia. For one thing, the Spanish were after resources in the Americas, primarily gold, and the settlement of new colonies for religious freedom was not a part of their agenda. The men who came over were adventurers and entrepreneurs, not good burghers intent on living a spiritual life in a New Jerusalem, and these adventurers treated the land and the native people in an adventurous manner, to say the least. The best thing that can be said about them is that, unlike the English and to some extent the French, they tended to intermarry with the natives—*Conquistamos las Americas con la penga*—as my old, irreverent Spanish teacher used to say.

This produced in South America and Mexico a separate group with no allegiance to either ancestral heritage. The land was granted, as with the English, by the Crown, but the grants tended to be to individuals rather than to companies or groups of individuals. According to Spanish law, the colonists not only owned the land, they also owned outright the native people who were already living on the land. The system, known as the *encomienda,* was a variation on an old Castilian legal doctrine that allowed the governor to commend the Indians to the colonists to use as needed and to instruct in the Christian faith. The grantee was responsible for his Indians; he had to feed and clothe and educate them. But he could also put them to work. The roots of this lay buried in the feudalistic soils of Spain, which in the sixteenth century were much deeper and stronger than the feudal manorial systems of England in the seventeenth century. The *encomienda* was, in effect, a manor, with the adventurous, often renegade, warlike lord at the top, the Spanish henchmen, also ruthless, in the middle, and beneath them whole nations of native serfs, or *peones,* as they were called. The Roman Catholic church was an integral part of this system. Priests often came over with the early explorers and colonialists and, although they accepted outright the system that enslaved a whole continent, they took it upon themselves to save the lost souls of the Indians. (The English, in contrast, had a running debate in some quarters as to whether the heathens were human.) The Church itself owned vast tracts of land and established colonies of converted Indians, not unlike the town at Nashobah, the only difference being that at Nashobah the Indians were in control of their own land. The Nashobahs owned the land within their town and could, with permission of the General Court, buy or sell at will. In the Spanish colonies, land was held by the Church and the converted Catholic Indians had no rights. The Church became a major conservative force in Spanish America, and when the revolutions finally came, as was perhaps inevitable, in places such as Mexico, the rebellions were as much against the Church as against Spain. The first thing the Indians did was kill the priests.

As in the north, the Indians in Central and South America had a strong concept of territory, and, as was also the case with most of the American tribal people, the land was thoroughly integrated with the culture in a sacred way as well as a political one. It was impossible to separate the two in the native cosmology—any split between matter and spirit was probably incomprehensible to them. The land was their way of fixing a place in the world, a way of anchoring themselves. It is a concept that is difficult for those of us who come out of a tradition that looks upon land as a resource, as something to be used, to fully understand or feel. The loss of a land base was spiritually and economically devastating to them.

Land surveys in the early Spanish colonial system were sometimes lacking or if they were made at all were very poorly done and covered only the tracts granted by the Crown. Records were poor, and there was no way to get more land or new land other than by a royal grant. As a result land titles in Latin America today are frequently defective and the old system prevails. Property is still held in large estates, the *latifundias* or the *haciendas,* and the landowner may not even be present, so that the system functions by a kind of tenancy. These tenants are wage workers who are paid either in cash or in kind, and who are often kept in debt in a state of desperate poverty on the *haciendas,* the veritable recipe for smoldering revolutions. Since the tracts under this system are often large and relatively unmanned, squatters are common, and although a family may live on the land for generations, they have no rights to it, no title, and so are reluctant to improve it. If the land wears out, the squatters move on. With an increasing population, this custom of squatters and freeholders moving onto new lands and clearing native tropical vegetation to make new fields is one of the primary causes of the destruction of rain forests in the Western Hemisphere.

In some sections of North America, particularly in frontier areas such as Maine, the same system prevailed. Early land speculators, known as great proprietors, held large tracts that they left untouched. In the late seventeenth and early eighteenth centuries, squatters moved in, cleared a tract of land, planted, and put down roots, presuming they were on *terra nullius.* The first sign of the presence of a landowner, a lord so to speak, was the appearance of the surveyors sent out to run the lines of the properties. These men came to be hated and feared by the early settlers; some of the surveyors were roughly treated, threatened, and one or two were even killed. Periodically in the colonial period, there was resistance to these great proprietors. In Maine a group of squatters who attired themselves in native dress and called

themselves the White Indians pressed for land reforms. (Why is it, I wonder, that so many reformers, from the Boston Tea Party to the sixties, feel it
necessary to attire themselves in traditional Indian costumes?)

There also were periodic land revolts by smallholders in New Jersey,
New York, South Carolina, and Georgia. One of the stiffest and most successful resistances came from a group living on the Van Rensselaer holdings
on the upper Hudson.

In North America, the feudal characteristics of tying land to a single family has never had much sway, but the system came close to reality in the
Dutch colonies along the Hudson River. In 1629 the Dutch West India
Company drew up a charter that allowed any member to buy vast tracts of
land consisting of sixteen miles along the shores of navigable rivers and running inland as far as the political situation allowed. The *patroon,* as the landholder was termed, had to establish a colony of at least fifty people within
four years of purchase and was the absolute administrator of the territory.
The most successful patroonship, and the largest (and last) feudal estate in
North America, was created by the Van Rensselaer family, which by 1637
held nearly seven hundred square miles on either side of the Hudson.
Tenants had to pay rent amounting to some two hundred dollars a year,
had to give three days' service to the lord, keep up the roads, and cut and
deliver firewood, wheat, waterfowl, and butter to the manor house. The
original proprietor was Lord Kiliaen and vessels moving up and down the
Hudson were required to strike colors when passing his manor.

After the Revolution the vast feudal estate of the Van Rensselaer family
was still intact. The current lord, one Stephen van Rensselaer III, known as
the Good Patroon, arranged to keep the grants he held by working with
Alexander Hamilton to devise a legal strategy to continue as he had under
the Dutch and the British. The two of them drew up deeds that conveyed
the lands to the farmers forever but still required yearly rents to be paid in
wheat and fowl and service. The land could not be sold without first offering
it to the patroon, who even after it was sold would receive one-quarter of the
price. Rents were exacted in full, many evictions took place, and there were
many court battles, but inasmuch as power back then—as now—lay with the
moneyed, the farmers lost again and again in the courts. So they revolted.

Under the leadership of an elusive figure known as Big Thunder, the
rebels dressed in their version of native dress, calico, and carried tin horns,
which they would sound to warn each other of the arrival of the rent col-
lectors. They grew increasingly violent, but they continued to work through

the ballot and in time became enough of a political force effectively to wear down the patroon. Over the years the manor was broken up by discharging the rent for relatively small cash payments.

France, the other great contender for North America, was also primarily interested in trade, mainly in furs. By 1608 Champlain attempted to create a colonial settlement at Quebec, which struggled along until Champlain's death in 1635, when his colony still had only about eighty-five residents. As in Spain, Catholic missionaries set out into the wilderness to convert the heathen, and by midcentury the Jesuits had established a mission at Georgian Bay, where they hoped to create a Christian Indian community. But they had failed to take into account the ferocity of the Iroquois nation, the same group that would periodically terrorize the gentle coastal tribes that eventually populated the village at Nashobah.

During this same period, Cardinal Richelieu founded a company, known as the One Hundred Associates, that was granted a colony in the St. Lawrence River Valley and had complete control over the fur trade in the region. The company made large grants of land to individuals, known as seigneurs, on the condition that they settle them with tenants. But this group, too, failed to bring out colonists in any significant numbers, and so the king created a royal province in the area, complete with a governor, a bishop, and a military commander with a regiment of soldiers. This group finally managed, in 1666, to defeat the Iroquois, and settlers began to arrive, including one of the key elements of any successful colonial venture—women of marriageable age.

In the end, of course, as we know all so well, England won North America. We speak the English language, we eat English food (unfortunately, the Solicitor says), and more to the point for this essay, we have inherited the system of English law (not so unfortunate, according to the Solicitor). It is also true that, even in sections of the Americas that were always under control of the British, the land-settlement patterns were different. In New England, generally speaking, a group applied to the king to form a settlement. After obtaining permission, the settlers moved as a group to the site and built a new town. Founding families of a town formed a covenant or corporation that held title to the surrounding lands. They carefully surveyed the lands, maintained title records, and laid the village out with a small common grazing area and a common or green around the meetinghouse, where individually owned lots with larger houses were arranged.   117

Working together, they cleared the land outward from this center. Land titles in this system were very clear, land speculation was at a minimum, and the settlers were homogeneous and generally of a single religious sect.

Colonial settlements in the South made use of what are known as head-rights. Each individual was given a tract of land of about fifty acres, from which he and his family were expected to garner a living. The system allowed for speculation. A ship captain or a planter who imported other colonists by paying their fare, or through the indentured servant system, was given the headrights of these indentured individuals. You could locate your land wherever you chose if the land was not already claimed, and although land was supposed to be surveyed and the deeds recorded, the surveys, as in the Spanish system, were not always accurate and the records not well kept. Cases arose in which more than one party claimed ownership of a tract. Speculation was common under this system and became a means of amassing a fortune, which is one reason that, in the American South, large plantations, owned by a single family, managed to develop.

As the English control of North America spread west, the southern system of individual occupancy was more the norm. The American image of the hardy individualistic frontiersman with his ax and rifle carving out a smallholding in the midst of a wide and dangerous land has evolved into one of the powerful American metaphors. Also one of the more dangerous. The idea of no government, of freedom, of no controls, of dog eating dog and every man for himself, if it did not exist already in the human soul, was certainly nurtured in the wild freedom of the open lands of the Americas. Many of these early frontiersmen settled illegally on the frontier. In the beginning of the colonial period, the occupancies were considered trespass, as they would have been in England. Laws and edicts prohibiting settlement on land you didn't own were passed, but over time these hardy frontier souls began to be favored. Those in the East looked upon the frontiersmen as a buffer between an even more dangerous notion, the wildness of the Indians who lived—now, finally—beyond the western frontier.

Wilderness, in the biblical sense, came to an end in America on May 20, 1785. On that day Congress authorized the survey of the Western Territories and divided the continent into a system of six-mile-square townships. Each township was to be organized by straight lines running due north and south, and due east and west. Within this boundary, there was a further division of the six-mile-square tract into thirty-six square sections, each amounting to 640 acres. This pattern determined the spatial organization of the continent outside the original colonial settlements. It was a rational, mathematical layout, an outgrowth of the Enlightenment, an attempt to bring order out of the chaos of wilderness. It did not allow for subtleties of landscape, for river valleys, ridges and hills, and the vast, geologic upwellings of the Rocky Mountains. No Mississippi, no Colorado, no recognition of the sinkhole of Death Valley. No place names. Anyone who has ever flown east to west across the United States on a clear day can see it there in the squared-off regular settlements of farms and towns west of Saint Louis, east of the Sierras, and north of the former Spanish territories.

The national grid was essentially the ultimate expression of private property, a way of laying out the whole continent into identifiable, salable lots. The system, by the turn of the nineteenth century, had the backing of American law, which viewed land not as a place of wildness and beauty or a source of sustenance, but above all as a commodity to be bought and sold primarily for capital gain. By the 1820s new settlers, including uprooted, dissatisfied New Englanders and newly arrived immigrants from Europe, could purchase land for as little as $1.25 an acre. The result was that even a poor person could own, by European and even New England standards, a vast manorial estate, a huge chunk of land. But this grid, this division of the world into squares, dictated a certain attitude toward land, a commitment to the private holding rather than to the village or the community as a whole. What lay beyond your holding was another world; your manorial lands would supply all in this theoretical paradigm. And so now, what lies beyond the uniform blocks of American towns, unadorned with public squares, central meetinghouses, churches, and cafes, is the infamous *strip,* a free-for-all no-man's-land stretching off into the outlands, where the greediest can, with abandon, erect the largest signs, serve the fastest foods, and offer the most convenient parking.

Maybe I am too Eurocentric, but I actually find this beloved vernacular landscape of fast-food architecture and the automobile downright terrifying. I'd rather take my chances with the sucking fens of Nashobah and the grasping tentacles of Ap'cinic.

Sometimes, driving the arrow-straight roads of the Midwest or flying over the United States, staring down at the squared-off, straight-running roads and highways and the regularity of the fields and town blocks of the grid, and knowing what's down there on those roads, I feel deep communion with Ratty and Mole in the *Wind in the Willows,* who lived in a world of quiet country lanes, of tilled fields and hedgerows, with a river running through and water meadows stretching back to a mysterious wildwood.

Chapter Nine

As he hurried along, eagerly anticipating the moment when he would be at home again among the things that he knew and liked, the Mole saw clearly that he was an animal of tilled field and hedgerow, linked to the ploughed furrow, the frequented pasture, the lane of evening lingerings, the cultivated garden plot. For others the asperities, the stubborn endurance, or the clash of actual conflict, that went with Nature in the rough; he must be wise, must keep to the pleasant places where his lines were laid and which held adventure enough, in their way, to last a lifetime.

–*Kenneth Grahame,* The Wind in the Willows

Holding Ground

I once visited the Solicitor's father, a stalwart old gentleman with the same sea blue eyes as his son. The family had once made a name for itself manufacturing automobiles in the time when cars in England were considered so dangerous they had to be preceded by a man waving a red flag. The old man had retired to a country cottage in Kent, not far from Tunbridge Wells, and since I was nearby, I paid him a call.

He lived in a section of fruit orchards, at the very edge of a pocket of wooded land that was commonly held—interestingly enough—by the property owners of the houses that surrounded it. The Solicitor's father and his neighbors shared the work of maintaining the woods.

I think the spirit of his automobile-loving family was still rampant in the old man's genes. Even though he was himself in his eighties, he was a driver for the English equivalent of Meals on Wheels, and furthermore he insisted on spending the day driving me around the countryside, showing me the sights. We went to the gardens of Sissinghurst—which were closed that day—and we went down to the battlefields and abbey at Hastings, where William the Norman defeated Harald and thereby solidified the European feudal system throughout the land.

Except for the fact that the architecture is different, and there are no castles, Kent is reminiscent of the nineteenth-century New England countryside—rolling, open lands, fruiting orchards, a few woodlots, and clustered villages. This makes perfect sense, of course. As I said, the county was the homeland of many of the dissatisfied Puritans who came to the New World and was the inspiration for the fruit-growing industry that still flourishes in the Nashobah Valley. It was also the region in which the tradition of free-holding of land was not unknown. The custom must have planted the seeds of the idea of private ownership of land, which, along with the apples, flourished so fruitfully in the inland communities in New England. Kent is also the native county of Wat Tyler, leader of the fourteenth-century Peasants' Revolt, the first great popular rebellion in English history.

Wat, the tiler, who lived in the town of Dartford, became enraged when a tax collector invaded his cottage and abused his daughter. Tyler struck the man dead, and the townspeople gathered around in support. One thing led to another, and the people of Kent joined with similarly dissatisfied peasants from Essex who were also protesting the unpopular poll tax that had been instated some years earlier. With Tyler as leader, the group marched on London, freeing prisoners along the way, shouting for the abolishment of property, and proclaiming all men equal. They reached Blackheath and camped there, and then the next day the whole mass of them marched to London Bridge.

When he heard word of their approach, the mayor of London, William Walworth, caused the drawbridge to be raised to prevent their entry into the city, but the citizens lowered it again and the peasants poured into the town, opening prisons and burning books and documents at the Temple and Lambeth Palace. They began breaking into wine cellars at Savoy and other manors, but in spite of the apparent madness of the crowd, Tyler held them in control, and there was very little looting; in fact, legend holds that they drowned one of their own party for stealing a silver goblet. What they wanted was a meeting with the new king, Richard II, who was only sixteen at the time.

The king and his entourage at first retreated to the Tower, but then agreed to meet the rioters the next day at Mile End. There were now some sixty thousand peasants collected together, and they met, peacefully, with the king on the following day with their demands. They had four conditions: abolition of serfdom, liberty to buy and sell like freemen, pardon for past offenses, and finally, and perhaps most radically, that rents on land be fixed at a certain price in money instead of being paid in service.

The king agreed, and kept his scribes up all night writing out a charter.

Tyler, however, was not at the Mile End meeting. He had remained in London to search out enemies in the Tower, and he had more conditions:

he wanted the forest laws, which prevented peasants from hunting and trapping and cutting wood, entirely abolished. Wat met the king in the streets in London, and there was some confusion—possibly a misreading of body language, possibly an outright threat to the king, whereupon Mayor Walworth, who was with the group, drew his short sword and stabbed Tyler in the throat.

The peasants assembled for revenge, but the young king rode out openly among them and declared that he would lead them in their own revolt. The whole troop marched on to Islington, where there was a large body of soldiers loyal to the king. Once he was safe, the tables turned: the king withdrew his conditions, soldiers rode on the peasants, and the revolt was put down with the usual end. Fifteen hundred rioters were tried and executed, and by the mid-1380s the countryside was back to normal. But the seeds of the idea of liberty had been planted in the English mind.

The England of Ratty and Mole—the water meadows, the wildwood, and the village—is fast disappearing, and it was probably always a myth, even before Kenneth Grahame used it as background for his gentlemen animals. But at least the idea was pleasant and it is a dream that still lingers in the American soul, even in the midst of the free-for-all development patterns of the lands beyond the cities. The only difference is that now, to preserve even a few remnant acres of this mythic landscape, you have to fight the tide. Linda Cantillon, whose common field of wildflowers was threatened with a housing tract, learned that early in her campaign.

Around Nashobah anyone who wants to build, to cite but one regulation, has to prove that the septic system for the house will work. To do this, one must hire a backhoe to dig a hole in the ground and then call a board of health inspector to come to the site and time and measure the drainage. If all goes well and the water disappears from the hole within the required time period, the builder may proceed with construction of the septic system, following, of course, the approval of the septic design by the board. And not just one board. There is a regional Nashobah Board of Health and also a local town board of health, and the conditions of both of these must be met before construction of the system. Both of these boards are subject to the regulations of a state board, and the whole process is subject to conditions promulgated by the federal government. This is necessary to construct the septic system alone. Other boards oversee the other elements of the so-called site work, the preparation of the land for building. None of this has anything to do with the actual structure, which, once the site regulations are met, will be subjected to another series of building, plumbing, and electrical codes.

"We ended up questioning this whole development because of regulations," Linda said. "I mean there were many proposed violations in the plans, so many things this jerk wanted to do with the place, like his big plans for a swimming beach. You can't just go down to the shore with a bulldozer, rip up all those plants, I mean I've found turtlehead down there, for God's sake, and arethusa. You can't just go in and tear all that out and then dump a load of sand in the water, kill all those fish or tadpoles. You've got to get a permit to kill things in this country. This guy wants to put in a mini-sewage-treatment-plant because the land is either too wet or too filled with ledge to get his monster trophy houses in there under the normal septic codes. The guy's a jerk. Believe me."

One summer afternoon, working in her weed garden, Linda saw a huge Lincoln Continental drive down her small road and park at the dead end near the trail that she had used to walk to the tract with her children until she was forbidden to trespass there. Inasmuch as a golden Lincoln Continental was not a common sight on her tiny street, she went down to see what was up.

"There he was, Fatso himself. Big cigar, right? Big yachtlike car. Radios. These guys in their shiny shoes, walking through the muddy trail. I couldn't believe it. These are the people who take apart the world."

It turned out that Mr. Developer was not a bad sort once he got to know Linda. "He was always polite," she said later. "He was a gentlemen. Not like some of the other nerds around here. At least he fought fairly."

He also admitted, privately, that it was a beautiful piece of land. And he said further, and privately, that he had perhaps taken on a bigger expense than he had intended, no small part of which was due to the fact that he had stirred up, in the process of planning this development, the hornet's nest of the Friends of Open Space.

The group was growing now almost daily, and had attracted to its body not only local people in the town who were anxious to save the property, but also a few outside advisors who knew a thing or two about the law. At one point, the Solicitor himself deigned to provide advice.

"They had a good case," he said. "They had a lot of local support. I told them they ought to go for the throat. Fee simple. Just buy it outright. It's not like the landowners wanted to destroy the place. They'd sell to the highest bidder. And from what I can tell, they had the town behind them. Paranoid group, though. They don't trust any official anymore."

The Solicitor winked over his pint of stout.

"That's a good thing, don't you know. I daresay they'll win their case."

The Friends of Open Space considered the Solicitor's advice—which by this time they had also considered on their own. They approached the landholders and found that it was true: they would indeed sell to the conservation group if the developer pulled out of the deal and the price were right. Price, of course, in a time of budgetary constraints (or imagined budgetary constraints), was no small matter. And in any case, at that point the developer still held the land.

All of this, in the mind of any developer, is strictly business. Land is not a geological fabric that makes up the material of a living ecological tapestry consisting of soil organisms, which, in the course of their short lives, break down organic matter and feed the roots of plants and complex, soil-dwelling creatures of a higher order, such as worms and sow bugs, which in their turn feed salamanders and red-bellied snakes, and shrews and robins, which in their turn are fed upon by snakes and raccoons and possums and hawks, who in their turn will die and fall to earth to be decomposed and recycled once more into the great turning pattern of life and death that is the essence of this ancient, curious, perhaps unique, and long-suffering system of life on earth. Land to a developer is not a biological system. Ultimately, land is not even landscape, the roll of hills, the pattern of trees along a ridge, a sweep of meadow, a wooded hillside above a lake, a setting of seasonal changes, of red buds and bursting leaves, the charge of wind in summer, the turning leaf, the falling leaf, the snows of winter, gray ice, a brown wall of trees, and then, in time, the reddish glow of the maple buds. Land, to the developer, is a speculative matter, a question of the balanced bottom line, an investment. If, for whatever reason, that investment proves questionable, if it appears that the amount of money put into a property will not, in the end, equal the amount of money extracted from that same property, then the smart developer, the sharp land speculator, will cease to be involved in that transaction and remove himself from the equation. Those who work in the field of land protection know this fact and will enter into a fight to save a property with it in mind. In effect, those in opposition to a given development are a part of the mathematical equation that will determine whether or not the investment is worth it. In the case of the Frost 127

Whitcomb land, the equilibrium began to tip, and finally tilted far enough to dump one of the contenders.

In March, six months after the plans were announced, our man from Lynn decided to pull out.

I took a walk on the land in question later that spring, and made a large loop from Long Lake over to Fort Pond, then around the south end of the pond and over the hills and down through Morrison's orchards to the Nagog Hill Road on the east side of the Morrison land. As I climbed the hill below his house, seemingly out of nowhere, Black Jack the Bull suddenly appeared in front of me.

He was no more than thirty yards ahead of me, and he came forward at such a directed pace that I thought that, by means of some technological surveillance device, he had determined that I had been continuously trespassing on his property and now, having caught me red-handed, meant to teach me a lesson. I walked slowly forward, trying not to meet his eyes. His stare was palpable, but I avoided it until we were in the act of passing one another—he nearly in the middle of the road—his road—and I off to the side like a good peon. I should have removed my hat and held it to my chest as he passed, but instead I looked up at him, as if I had not noticed him before, and said with a slight brogue, "Good afternoon to you, sir."

He drew up slightly—very slightly—and I think I heard a sound emerge from deep within his massy chest, a grunt, a distracted piglike sound that acknowledged my presence but little more.

I should say I was better off than some sojourners in this place. One friend of mine, an eighty-year-old contemporary of Morrison's who, like Junior Kimball, rides his bicycle around the town for exercise, had passed the lord in this very spot some months earlier. These two octogenarians had had a recent difference of opinion over some town board matter, and Morrison had stormed out of the room. On this day he was standing with his manager when my friend with the bicycle rolled past and greeted him in a friendly manner, unbegrudgingly. Raging Bull met him with a blank stare and then turned to his manager. "Do we know this son of a bitch?" he asked.

Another neighbor was ridiculed by this bull for staring. The neighbor saw what he presumed to be a disheveled, homeless man standing under the apple trees above the Morrison house one summer afternoon as he was walking by. He couldn't help but look at the figure, repeatedly, I suppose, and suddenly this tattered man beneath the trees jutted out his lower jaw and pounded his chest gorilla-fashion, as if to say, "you want something to look at?"

At least Morrison was holding his own against all comers, friend and foe alike. At least he was taking good care of the property the law indicated was his own. He could have sold out to a developer. But that was not his way—any good feudal lord defends and even expands his fiefdom.

The same could not be said for other landholders in the region. Over at Beaver Brook, where the former farm had been sold off to a local developer, defense of the land and its noneconomic values was falling into the hands of those who merely used the land, not those who technically owned it. During the same period that Linda was fighting to save her tract of land, neighbors in the Beaver Brook area, most of whom were newcomers who had settled there in the 1960s, were building a campaign to control the development in their area.

This piece of property, which had become, de facto, their walking grounds, consisted of some 160 acres and up until the 1970s had had a barn, a large chicken house, and three separate hayfields. The farmer who owned the property, a dairyman from a town to the north, used the land to cut hay and once grazed his heifers there. But old age caught up with him, as it will do. He removed his heifers, and came later and later in the growing season to cut his hay, until finally he came no more. Queen Anne's lace and yarrow and daisies and black-eyed Susans spread amidst the pasture grasses. The black birch and the red osier dogwood jumped in as soon as they had a chance; the woodchuck burrowed into the deep sandy soils along the Beaver Brook; foxes and coyotes patrolled the banks; and slowly the fields began to devolve toward a natural state, which, in this part of the world, consists of deep forests of oak, hickory, and maple.

This return to wild land may not have been good for agriculture, but it was good for the children of the neighborhood. Here, in the thirty years since they had lived in the area, the children and the adults had roamed through the woods and fields, cross-country skied in winter, and fished from the old narrow bridge over Beaver Brook.

Mainly, the people in the area were worried about the loss of peace and quiet as a result of the development—not an uncommon concern, it turns out. A few years ago, the National Real Estate Board did a survey that found that what people most seek out in a new house is peace and quiet. Second on the list is nearby open space, green land. Golf and tennis and gyms, often the most advertised aspects of a new development, are low priorities.

One summer evening, in order to understand the opposition, the neighbors invited the developer himself to come and give a presentation at one of their regular meetings. This particular developer was a man who was not appreciated in the town of Westford. He had undone many a beautiful parcel with his handiworks, one of them on the sloping eastern banks of Beaver Brook where, on an old abandoned farm and a rich wooded slope, he had constructed a housing tract that, in a wild flight of imagination, he had named Hitching Post Common. I knew the tract well before it was destroyed, and there was not a single hitching post in the entire three hundred acres, and whatever was left of the common grazing lands had been effectively obliterated a century before when the farmhouse in this area, originally owned by David Flagg's family from the Great Road, had been privatized.

Said developer had gotten very rich through his works, richer than was deemed necessary by his opponents. The big man himself did not appear at the meeting. He sent his front men, two perfect gentlemen in short-sleeved shirts, neatly pressed Dockers, and tasseled loafers. They came with expansive maps and design sheets and prepared presentations that described, in so many words, an earthly paradise of houses. Actually they gave two descriptions. One was a standard subdivision with curving roads and house lots lined up one after the other in the boring, unimaginative, deadening pattern that characterizes the current landscape of America. The second was a cluster development in which the houses were crunched together in smaller lots with much open space in between and a common area to one side. (No one seemed to notice that the common area was already destroyed by a gravel operation and that the open space, valuable though it may have been, was wetland that could not—without an elaborate permitting process—have been developed in any case.) Essentially, the plan was a bone the developer was throwing out to satisfy what he must have recognized as a pack of territorial wolves.

The wolves liked the cluster idea but did not bite on the bone of "common land." This section was already common land as far as they were concerned, their common. Never mind that it was privately owned—they had been using it as a common resource for nearly a generation. Children who were now marrying and starting careers had played there. The tract had become for them part of their collective memories. But at this stage of the process there was only one recourse for the neighbors, and that was to hold this development at bay through the permitting process, as the Friends of Open Space were doing over by Nashobah.

I went to one of these permitting meetings to follow the course of events at the Beaver Brook development. The meeting began at some ungodly early hour, seven o'clock or so of a fine summer evening, and was held in the cellar of the local library. It was packed with people. Good, I thought, great turnout. Then I realized that the development the assembled were discussing, which had generated some heated debate, was not the one I had come to observe. Other developments in the town were being subjected to the same strict review by other neighbors. You got a sense there of the passions associated with change. The only dispassionate players that night were the board members, who went through this tedious process week after week, and the developers, for whom this type of thing was business as usual.

There was an ominous group of men in dark suits lurking in the back of the room at the meeting, and when the Beaver Brook project was called to the floor, these men advanced in a gruesome phalanx, armed with maps, briefs, and plans. They were the attorneys for the developer, it turned out. Much legal jargon, much displaying of charts and diagrams and plans, many tedious words, and many questions from the assembled board members. Within ten minutes, the plans were sent back. Papers missing, deadlines unmet. It seems that, among other sins, the developer had submitted his plans after a moratorium had been called on new developments in the town. Project denied.

Inasmuch as he was a man of experience, who knew a thing or two about the law, the developer proceeded apace, as developers will, and sued the planning board. The proposal for the Beaver Brook project continued, suits, denials, animosity, and opposition from the neighbors notwithstanding.

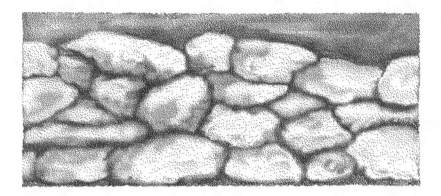

Chapter Ten

They are of complexion like our English gypsies, no hair or very little on their faces, on their heads long hair to their shoulders, only cut before, some trussed up before with a feather, broad-wise, like a fan, another a fox tail hanging out. . . . They made semblance unto us of friendship and amity; they sang and danced after their manner. . . . Some of them had their faces painted black, from the forehead to the chin, four or five fingers broad; others after fashions as they liked.

–Mourt's Relation: A Journal of the Pilgrims
    at Plymouth, *1621*

Out of the Quiver
of the Scriptures

I once stumbled upon one of the revivals of the old Indian powwows and dances that occur even in New England, where Indians were extirpated—mostly—some three hundred years ago. I was aimlessly driving around the area north of the Nashobah tract, thinking, as usual, about Sarah Doublet, and saw a sign that said Indian Powwow, with an arrow, so I followed it to an old 4-H camp set in the woods and then followed the sound of drums to the big event. Here, in an open riding ring under the trees, I came upon a vision—not of ancient native practices and rituals but of modern America.

A ragged band of people of all ages, dressed in cowboy hats, deerskins, trade cloth, and feathers, was shuffling around the ring to the steady beat of a deep-voiced drum. Overweight men in black cowboy hats with silver bands sat around the great skin drum pounding out a steady rhythm. In the ring, each circling in his or her own world, were buffalo men in robes with horns affixed to their heads. Here came an old tracking man, his bowed legs carrying him forward step by step, his black eyes searching the ground for some imaginary animal track, head snapping this way and that as he searched. Here was a Sioux in full regalia, shuffling, then leaping in the air with his ceremonial feathered spear, then shuffling forward again. Some merely pushed along chatting to one another, some twisted their bodies over into bizarre contorted poses and stamped hard upon the ground, and some carried spears or rattles or shields decorated with the skins of foxes and coyotes and bears.

Some participants in this dance were clearly Native American, with black eyes and copper skin and long jet black hair, worn by both men and women in ponytails. Some were clearly white, or had a generous dose of European blood. Many of the dancers were dressed like cowboys, in jeans, wide-brimmed hats, and boots, and some, including the whites, were attired in full Indian regalia. There were Plains Indians with feather headdresses, Abenaki Indians in blue trade cloth trimmed with red, Massachusetts and Pawtuckets—Sarah's people—all in subdued browns with grouse-feather headdresses, deerskin leggings or skirts, and strings of wampum and trade beads. There were white men with blue eyes dressed in loincloths with arm-bands of silver and leggings of deerskin; there was a beautiful Indian princess in a green fringed dress with loose black hair and a feather that she waved to and fro as she circled. Here came a huge blond Viking in horns and silver necklaces (perhaps, I mused, he had been captured by the dreaded *skraelings,* or local Indians, who struck such fear into the Vikings, and had been brought south to Massachusetts as a captive). Following him was Queen Bluebird, a woman of thirty-five or so dressed in blue feathers and a blue skirt. Mixed in with these exotics, and in some ways most exotic of all, were ordinary white suburbanites, some in their peaked tractor caps and baseball caps, fresh jeans, and pressed flannel shirts, blue-haired housewives with perms and clear plastic glasses with silver and gold flecks embedded in the frames. There were many children, and many young girls in regalia, some dragged to the event by their parents, but some clearly enjoying the dance. Two or three little girls of about ten were circling together with fancy steps and rattles. And there was one poor sensible little boy of ten or twelve who shuffled along behind his ecstatic father with a deep frown, clearly hating every minute of this absurd event.

For what seemed like a full hour, the vast congregations shuffled around the ring in a mass of humanity, their eyes glazed, having deserted perhaps, for the first time in weeks, all thoughts of daily life, which I suppose was the point.

For their predecessors, the real Indians, the dance was also escape. But it was also necessary; it was a communion with the universe, a means of binding themselves with the commonly held land that supported them. Without it, the world might fall apart. The English recognized that danc-ing was critical to the Indians' spiritual life, and they must have recognized the fact that dance was closely related to fertility rites—after all, they came from a country where morris dancing was still common. And so, when Eliot drew up the contracts for the converts in his Christian village, he outlawed "pawwawing." It was a way of breaking the culture.

There was much to break. Another was the native concept of work. From the English point of view, the women were carrying the burden of daily life in the Indian community. They carried the children, they foraged for plants, they worked the gardens, they built the houses, they cooked and cleaned. The men by contrast would go out on hunting forays, they would fish, and they would make war. In between they would remain in the villages, fashioning weapons, lounging, passing the pipe at ceremonial times, and gossiping. Renegade Anglican upper-class colonists such as Thomas Morton quite approved of this Indian way of life—hunting and fishing at leisure, smoking and relaxing in between. He especially enjoyed the Indian custom of changing residences according to the season; it reminded him of the summer and winter houses of the nobility back home. But to the Puritan English, who, next to godlessness, detested sloth, the Indian way of life was abhorrent. One of the requisites of conversion to Christianity was that everyone would labor, including the men.

Members of the tribe, or confederation, to which Sarah most likely belonged, the Pawtucket people, were essentially agriculturalists who supplemented their diet with wild edible plants and game and even managed wild lands to improve berry gathering, deer hunting, and fishing. The archaeological evidence is still unclear, but agricultural practices may have come to New England from the Midwest, where farming had been a standard way of life for as many as ten thousand years. By the 1500s, the so-called Contact Period, when Europeans first began visiting the shores of North America, farming was an established way of life among the southern New England Indians, less so among the native people to the north.

Most of the agricultural work was under the charge of the women. By stripping bark to girdle trees and burning the standing trunks, or in some cases by felling the trees altogether and burning them, the women would clear a given area to allow the sun into the garden space and then work between the stumps or dead trees. Using hoes of sharpened sticks, or moose shoulder bones or clam shells in coastal areas, they would break up the heavy sod and rake it into hills, each a foot or so across, and spaced about three feet apart. The band or family group would use the field over a period of years or for as long as the fertility lasted, and then move on and clear another 137 area. There was plenty of land and few people. This was essentially swidden agriculture, and had the English not arrived, it is probable that Indians would still be living here in the old manner, planting under the old traditions—after all, three hundred years is not a very long time in their history. It was in effect what modern day agronomists are ever striving for, a sustainable system.

In established fields in early spring, the women would rake together the detritus from last year's garden, the weed stalks and the stems of squash and beans, and burn it, using the ash to fertilize the hills. The actual planting season began in the Nashobah area after the last frosts had passed, generally about the same time that the oak trees begin putting out leaves—the old legend is that when the young leaves are as large as a mouse ear, it's time to plant.

Although there is currently debate in anthropological circles on this, it is also likely that they used fish, two or three to a corn hill, to fertilize the ground. Even as far inland as Nashobah, which lies some thirty-five miles from the coast, anadromous fish would have been abundant in the local streams each spring. Nashobah Brook is only half a mile or so from the village site, and the larger Beaver Brook is about two miles to the northwest. In the 1670s, Tom Doublet, Sarah's husband, maintained a fish weir there. He would have further narrowed the brook at an already narrow spot by using stones or brush and then constructed wicker traps upstream from the sluice where the migrating fish would collect. Then it was a simple matter of scooping them up to smoke for later consumption, or to use as fertilizer.

While the earth warmed and the fish rotted away, the women and the children, and maybe the men as well, would have had to tend to the business of keeping their dogs and the wolves, skunks, and foxes out of the planting grounds, away from the fish. In some areas they actually fenced the fields, but in most it simply meant keeping a constant vigil. Above some of the fields the Indians built wooden towers, and when they were not working the fields, they would post a boy in the tower to shout and clap and scare away the raiding scavengers and, later, once the seeds were in, the raiding flocks of birds. They also built what Roger Williams called little "watch houses" in the middle of their fields where these watchboys would actually live.

The Pawtuckets grew the traditional three sisters of Indian agriculture: corn, beans, and squash. The corn would provide a place for the beans to climb, and the beans and the corn together would shade, in full summer, the squashes. The women would poke holes into the hills either with their fingers or a planting stick and place a few seeds of corn along with a few of beans. Later, between the rows, they would plant the squash seeds. All summer, as the crops ripened, the women, some of them carrying the children on their backs, would work the gardens, keeping the weeds free from the young seedlings until the squashes grew large enough to shade out invading weeds. If growing conditions were good, by early July the corn would be about two feet tall, and the squashes and beans beginning to ramble. Dry weather, which in the Nashobah area is almost a given in the month of July, must have been a problem for them, but it is possible that the two ponds and a small spring on the west side of Nagog Hill may have been used to supply water to the planting fields.

As the corn grew the women would rake up soil around the hills and continue to weed. By late July they would have been able to pick fresh beans and young squash, and by August, the gardens would have been in full fruit. The corn ripened about this time, and before they became Christians, and no doubt even after they became Christians, they held ceremonies to celebrate the crops, especially the corn. In late August, when the corn was fully ripened, they would collect a pot of kernels and set it out as an offering. I suspect that as long as Eliot or his henchman, Daniel Gookin, was not around to instruct them otherwise, they may have offered a pot of corn to the Lord, and perhaps also to Jesus Christ, along with Keitan, the god of the west wind, and Manitou, the spirit force that dwells in all living beings and inanimate objects. Perhaps even a pot to appease the horned water beast in Nagog Pond, as well as Hobamacho, not to mention the devils, mouse spirits, the bear and the moose, the wood-dwelling tree spirits, and the demons and ghosts who lived and had their being in the great circle in which these people existed.

I have a friend who lived among the Naskapi and Montagnais of northern Canada who claims that Christianity for Sarah's people would have been a relatively easygoing religion. They could pray to one god to solve all their problems and didn't have to go around propitiating every living thing before they acted, asking forgiveness after each hunt, and making certain that the animals they killed would be happy in the afterworld. "Compared to animism, Christianity is an easy ride," he said.

More likely, though, with the advent of Christianity, they simply took yet another god or two into their pantheon. They must have understood that for some bizarre reason the teachers of Christianity did not like them to follow their old religion. But with the missionaries out of sight, they must have brought out the old skin drum from time to time, built a huge fire, and after sundown started a dance to celebrate the harvest, just as they and their people had done over the past two or three thousand years.

Indians of all cultures in North America danced. They danced for the harvest and the hunt, they danced for war, for death, for birth, for planting, for rain when there was not enough rain, and for sun when there was too much rain. Indian dance was, according to the cultural historian and art critic Vincent Scully, the greatest art form ever developed in the Americas; it was a symbolic, magical assurance, a way of keeping the universe functioning, of ensuring the continuation of existence. It was so ingrained, so deeply rooted in the culture that it is not likely that these late-coming English Christians, with their news of Christ the Redeemer, could have rooted it out altogether, even though it was specifically forbidden by the missionaries.

The Puritans did not like dance very much, or at least had problems with it whenever it involved couples. They were hard pressed to find biblical proscriptions against it, although Increase Mather did his best in a diatribe against the act he published in 1684 entitled *An Arrow Against Profane and Promiscuous Dancing Drawn from the Quiver of the Scriptures*. His arrow was mainly aimed at the vile, infamous, and abominable practice of "gynecanadrical dancing"—that is, men and women dancing together. Mather cited the Seventh Commandment and argued that it forbade the "Devil's Procession." Country dancing, the folk dances they had brought with them from England, and which have now resurfaced as contra dances, were not so frowned upon.

But Mather had the right idea. It was, as he wrote, a regular madness. Practitioners of this sin were followers of adulterous gods, he said, apostatizers, idolators, Israelites, and Greeks who worshipped Bacchus.

God forbid that he should have been at Nashobah on the night of the Green Corn Dance, among the circle of his good converted Christians, when things might have got out of hand, and the heat of August swelled below the hills and the great skin drum throbbed and all the recent converts slipped back into atavistic deviltry, and the shuffling, mumbling circle shifted blindly around the fire. In place of the Lord's Prayer, or perhaps along with it and other memorized words of the Scripture, the ancient vocables of their original culture came creeping out of the dark forest again and entered into their souls. Late in the night, after as many as eight to ten hours of dancing, all the graveyards on the northern slope of the hill yawned open and the ancestors emerged, and the waters of Nagog rolled back and

Ap'cinic came slithering up the hill with Hobamacho and Bear, and they would see them all there in the firelight glitter, their hideous wild faces all furred and toothed and their eyes moist and gleaming.

Dancing, as Increase Mather knew all too well, broke down the normal flux of time, broke through the barriers that lay between the temporal and spiritual worlds so that, in the minds of the people circling the fire, the two were indistinguishable. Now the forbidden powwows and shamans spit and bled and went into trances and shivered upon the ground. By dawn, when the sun rose over the forked hill to the east of Nagog, they were exhausted. But they would have cleansed themselves in the same way that tragedy cleanses, or the spirit of Christ.

The Indian system of agriculture was well suited to the native environment, even perhaps for the women, who, after the English introduced an alternative, never seem to have questioned their role in the Indian culture. In fact they had more political power than the English women. The natives had developed a sustainable ecological balance with the land, a more or less satisfactory compact with the earth that allowed them to endure. Unlike the English system, which required heavy physical labor, the Indians, including the women, had a lot of time to sit or work at smaller chores that were not physically tiring. They were none of them used to sustained, tiring labor, save perhaps the hunt. In fact, a few years ago an anthropologist did a study of the use of time among hunter-gatherers in the Australian outback, one of the most unforgiving environments on earth, and found that people spent most of their time staring into space and hunted and collected plants less than four hours a day all told.

The Puritan farming method, although aided by Indian traditions such as the use of corn hills, was essentially an English import, and as such required a total transformation of the land. The Puritans attempted to plant wheat and oats, neither of which did very well in the New England environment. Barley and peas did manage to survive, but the best crop, and the most reliable, was Indian corn. It was not a favored meal for the English, but it did well, and furthermore it could be fed to hogs and the stalks could be used for cattle fodder in winter. The English had also learned from the Indians how to use fish for fertilizer, but while the fish were rotting, they had the same problems with rummaging dogs that the Indians had. Farmers had to post people in the fields to keep animals away. The ground was spaded—in the early years there were not many plows in New England—and the grain was toed in. Once the seed was planted they had the same problems with blackbirds the Indians had, but rather than post children in camps in the fields, the English, with the bravura that characterized the whole Western attitude in the colonies, attempted to remake the world to fit their needs. To do so they relied on the law. The General Court *required* citizens to kill ten blackbirds and two crows a year. It was a simple but ecologically disastrous attitude.

Planting of the various crops meant clearing the land of everything, even the stumps, which, in the earliest fields, were lined along the edge as fences. Pale or picket fences surrounded common fields—the classic stone walls of New England were a later addition. The maintenance of these fences was a major consideration, as anyone who has raised a large garden near woodlands knows, even today. Hogs and cows were turned out to forage, and these animals, along with goats, would commonly break through to the planting fields to have their fill.

Early court cases involving trespass more often than not involve a suit of one party against the other because of damage by domestic animals. In order to maintain the fences, the office of fence viewer was created, a position that still exists today at Nashobah, but which is more or less a relic. Nowadays, once a year, the fence viewer drives around the town and checks the town boundary markers and is occasionally called upon to comment on boundary disputes, of which there are very few in the Nashobah area. In colonial times if your section of fence on the common land was not maintained, you could be fined.

The Puritan desire to induce the native people to spend their time working was due in part to a recognition that idle hands are dangerous. Sloth was, in the eyes of the government, not just a sin but, in the theocratic organization of this brave new world, illegal. You could be fined two shillings if you lingered over your beer in a tavern for more than an allotted period, usually an hour. If you were a repeat offender, you could be whipped or placed in the stocks, where, by way of entertainment, you might be pelted with rotten fruit or even rocks by celebratory local boys. Recalcitrant idlers would be whipped out of town, then whipped in the next town.

But there was another motive. Under the English system the Indians had to be fixed in one spot over a long period of time. Once they were settled they would be easier to control. For all the spiritual labors of this servant of

Christ, John Eliot, there was a practical political side to the business of encouraging people to settle down at Nashobah. Once settled you could be watched. And if by chance there was trouble, you could easily be rounded up and transported.

In the years before she was taken to Deer Island, Sarah would have been thoroughly immersed in the daily life at Nashobah, the rounds of planting and praying—and dancing—that made up her strange bicultural world. Every morning in the growing seasons of her youth, she would go to the fields to tend her crops. She would hill the gardens where she had previously buried herring, drop the corn seeds, and later the beans and squash seeds, into the hill, and go about the business of keeping the dogs and the birds away from the fish, and the seeds and then finally the young shoots. Most likely she would not have had to worry about livestock since the Indians at Nashobah probably did not keep any cows or pigs. Nashobah was still at the edge of the great wilderness, and so hunting in the mid-seventeenth century was still good. Furthermore, the runs of fish on Beaver Brook and Nashobah Brook were still strong. Nagog Pond, Fort Pond, and the nearby Long Lake, all of which were connected with a chain of small streams and wetlands, had a supply of "manie good fishes," as one early account explains. Eels, which are catadromous—they breed at sea and swim inland to spend their adult lives— are still found at Long Lake.

One of the earliest and newest crops at Nashobah would have been apples. The first governor of the Massachusetts Bay Colony, John Winthrop, was something of a gentleman farmer in England and an orchardist, and the first settler of the Boston neck, Blackstone, had planted apple trees at the site, so there were already apples growing when the Puritans arrived. The trees were probably Yellow Sweetings, a common English variety, but the settlers soon developed a local variety called Roxbury Russet, and there was a variety called Rhode Island Greening by the late seventeenth century. The Kentish farmers of the seventeenth century grew Pearmain, Russet, and Winesap at the time of colonization and probably brought these varieties over with them either as slips or as seeds.

It is likely that the Indians at Nashobah grew their trees from seeds and raised what are known as pippin apples, which would have been poor table apples and were used primarily for cider. These too would have been Sarah's responsibility. She would have planted the seeds, or the slips if she had them, and once they sprouted, she would have watched them, watered them if necessary, and, in one religion or another, perhaps prayed over them until   143 they were of sufficient strength to fend for themselves. Later, much later, she would have gathered the first fruits to press into cider. And this she would have allowed to ferment.

In his narrative of the Indians of New England, Daniel Gookin complains that the people of Nashobah, although pious, had problems with drunkenness. The English had taught them how to press cider, which in the seventeenth century—as it is today in England—was an alcoholic beverage. They may have even distilled the cider into a more powerful apple brandy.

With the crops up, Sarah would have gone about the business of weeding, then harvesting, then grinding, then drying and storing, and then gathering wild food. All of this was a hedge against the great scythe of life in her region, the New England winter. Indians would store the grain in baskets and over the season would have fed on a stew of meal supplemented with berries and wild fruits, to which in season they would add game that the men had either hunted or trapped. This included deer, moose, bear, and turkey, as well as smaller fare, such as rabbits and birds. Under a traditional system that was probably maintained at Nashobah even after the Indians became Christians, there would have been a sachem—a hunting chief or sagamore—who would have assigned to groups of individual hunters certain hunting territories or lines within the sixteen-square-mile tract, perhaps even beyond the tract into the traditional Pawtucket hunting territory. Prey might be killed individually, but the downed animal belonged to the group and was meant to be shared equally. Whether this always happened is questionable. There are records—albeit not in this region—of hunters butchering their prey out in the woods away from the village center so as not to have to share it all. A good hunt, like a good harvest, might have called for a celebration, another dance.

At the time of the harvest dance, the Green Corn Dance at Nashobah, John Thomas Good Man would have brought out cornstalks and gourds and perhaps even, by way of sacred object, a copy of Eliot's *Up Biblum God,* the holy written word translated into Algonquian by Eliot, which, according to Daniel Gookin, was in the possession of the people at Nashobah. Good Man may have joined the circle holding this magic symbol and dancing as intensely as the others, who waved the cornstalks and gourds through the high, smoky air, while around them the drums hammered and the great circle dance shuffled. The night would have been filled with the yelps and shouts and the long nasal vocables that characterized their singing—the "noyse" that so offended the Puritan ears.

It is possible, likely even, that Sarah and her people were drunk at this time, gloriously drunk. The Puritans—to their credit, some would say—may have been opposed to dancing, and fornication, and drama, but they were not opposed to drinking—not at least in moderation. In fact they were not opposed to anything in moderation exactly, although you would not suppose so from the fiery sermons and diatribes of Cotton Mather and company. But what they wrote and what they actually did were two different matters. The Puritans enjoyed healthy draughts of beer and ale and hard cider with their meals—there are even records of them complaining that in certain areas they were exploring for settlement, and also in the early years at Plymouth, there was nothing to drink but *water.*

But what they did among themselves they would not tolerate among the savages whom they hoped to convert. They had a system of fines at the Christian Indian villages for drunkenness, and Daniel Gookin expressed the desire, the hope, that he could undertake to halt the excessive consumption of cider that was going on among the people of Nashobah. By 1675, however, the issue would resolve itself.

In the early winter of 1675, Sarah was either visiting or living at the village called Wamesit to the north of Nashobah. Many of the people at Nashobah had close relatives or allied bands among the other Praying Indians, and Sarah had traveled there with her son, a boy of twelve.

By this date the Pokanoket leader King Philip had organized enough disparate bands to make an attempt at outright war on the English. There had been many abuses in the eyes of Philip, one of which was that he believed that the English had poisoned his brother while he was being held for questioning. But one of the worst offenses in Philip's eyes, and in the eyes of most of the allied tribes, was that the English were usurping the best Indian planting fields and then excluding the natives. Philip and his brother had been brooding on this for a number of years, and after much legwork, had convinced the tribes of New England, many of whom were traditional enemies of one another, to unite against the common enemy.

The war began in Rhode Island after a few skirmishes involving cattle killing, and soon the storm spread. The Indians of the Concord and Wachusett region, even those who were not Christianized, were a generally peaceful group who lived more in terror of the raids of the man-eating Mohawks to the north than the English. For their part, however, the English were unable to discriminate. Indians were Indians. Militias were formed in the smaller villages to carry out periodic raids against Indian villages. Sometimes these were merely groups of men, perhaps organized after an evening at the local tavern, who felt obliged to protect the territory.

One of these bands of renegade English from the town of Chelmsford organized a raid on the peaceful village at Wamesit. Early in the morning, while the Indians were still in their wigwams, an armed band of fourteen men appeared and called for the people to come out, which, having had business with the English, and furthermore having considered them more or less allies in this unpleasantness that was sweeping their territory, they did. When they were all outside, the English opened fire.

A great howl went up. There was a cloudy drift of black-powder smoke mingling with the ashes and woodsmoke that were such a part of the villages of this time; there were screams of the wounded and outrage at this monstrous betrayal, perhaps a few prayers, a few incantations. It is possible that the shock of the outburst caused some of the English to reconsider. Most of the people who had come out of the wigwams were women and children, and these now lay scattered on the earth, five of them writhing in pain. Sarah was there at the time, blood streaming from a wound, the histories do not say where. She had a worse injury though. On the ground, her twelve-year-old son lay dead. His name is lost to history, just one of the many thousands, some would say millions, of native people who had fallen and would fall beneath the fire of the Europeans in North and South America. The boy was the direct descendant of Tahattawan himself, the great Christian sachem of the blood.

No record tells us what transpired after this event. Sarah went back to Nashobah perhaps and sat by her fire and began to mourn in the old style, even though it was expressly prohibited by English law. She took up a freshwater mussel shell or a stone spear point and slashed her cheeks until the blood ran down around her shoulders, and smeared her face with ash and began the forbidden howling. She would have taken one of the shells or stone knives or perhaps, if she had one, an English steel blade and let out the long raven locks of her hair and twisted them into a shock, and then, still howling and crying and rocking by the fire, she would have cut off her hair, the shearing of locks among them being a sign of pain, a way of mourning the dead.

They must have wailed a great deal in those years at Nashobah. Unconverted warring Indians were circling the land at Nashobah, burning out the white villagers, killing and maiming, and taking the English hostage, and on the other side, the suspicious English had worked themselves up into such a frenzy of fear and revenge that they would just as soon have massacred all the Indians, Christian and otherwise, and be done with it. Sarah and her people must have mourned often as unconverted people they knew were killed or captured and sold as slaves. But Concord and Nashobah remained untouched during the early months of the war, until the autumn of 1675. Just about the time that the apples were ripened in the orchards, the war came home to Nashobah in the form of Captain Mosely.

Chapter Eleven

Trespass . . . Defendant has "stepped across." He has gone, unwelcome, onto plaintiff's property . . . an assault on neutral ground.

      –Charles Rembar, The Law of the Land, *1980*

It is admitted, that a fox is an animal *ferae naturae,* and that property in such animals is acquired by occupancy only . . .

      –Pierson v. Post, *1805*

The Last of
the Commons

It is possible that there is something ecstatic in the nature of the land in this place, something that induces human beings to celebrate, each in his or her own fashion. People who have no respect for the nature of private property come to this site by cover of darkness and lay here the ceremonial objects of their obscure religion. Neo-pagans have set up a celestial tracking stage here, in the form of a circle of stones, wherein they worship the rising or setting of the sun, moon, and stars. The descendants of the native people of the Northeast, as some claim, may be coming here at night to leave pine cones and bundles of brush as a way of honoring the place.

But other less holy people also come here. I have stumbled upon them myself.

One night in summer, coming back from a late event, I decided to swing through the tract to see if I could find any of the purported native people who come to the area after dark. Presumably, this being the twentieth century, these latter-day Indians drive to the site, and at one of the pull-offs near the bottom of Nagog Hill I saw a late-model Chevrolet, much battered. Ah ha, thought I, trespassers. Indians. Shamans, holy men, come to honor the sacred hill.

Up at the serpent mound, where I had often found these so-called brush donations, there was no sign of anyone, so I decided to try the shores of the pond, where I had also seen a few small stacks of brush and piles of stones (not to mention a few beer cans and the litter of illegal fishermen who also trespass on these sacred grounds). On the point of land now held by the Concord Water Department, which once had a big summer house at the site, there is a rocky shoreline that provides access to the forbidden waters. As I moved out onto the point, I heard quiet laughter, then a great splash and outflow of breath, and came upon three teenagers, two girls and a young man, skinny-dipping in Concord's drinking water.

As a fellow trespasser and objective observer, I felt this was none of my business, and turned to retreat and nearly crashed into a fourth teenager carrying a case of beer to the site.

"Sorry, man," he said after a torrent of surprised expletives.

"Don't worry, I'm not the police," I said.

Then I asked him if he came here often and if he had ever seen other people here at night on the other side of the road.

"Yeah," he said. "You see weirdos sometimes."

"Weirdos?"

"You know, like weirdos. Freaks."

By now the swimmers had spotted us, and having dressed or at least covered themselves, and having determined that I was not an authority, joined in.

"You know, really weird," one said.

"Like, you know, men with glasses. Geeks."

"Nerds?" I asked.

"No, not nerds exactly, geeks, like really weird."

One of the girls joined in enthusiastically.

"One night, we were here, and, like, there was this guy, and he, like, *looked at us,* and then he like went off into the woods, and he didn't say hello, or what are you guys doing here, or anything. He was so–like–*weird*."

"He looked at you?"

"Yeah, weirdly. It freaked me. You know, he just like–looked."

"I remember that guy," her boyfriend said. "A geek with glasses."

"Then there was this other guy," she said, "remember him? Some kind of dude with a box."

"That was a fisherman."

"No, no, there was a weirdo with a box, and he had like body parts in it that he was going to throw in the pond, so the cops couldn't find it."

"Body parts?" I asked.

"It was just some kind of fisherman, Tracy, it wasn't a guy with body parts."

"She always sees this stuff," another said.

"No, I could sense it. He was weird," she said. "It was his wife or something. You know, like that guy who sang in the church choir and then killed his wife."

"That was a long time ago, Trace. Don't listen to her, none of this is true."

"There was a guy in the choir, and he like killed his wife when she was out running, right near here, too. My parents knew him, and he was really weird, sang in the choir and went to church, then he kills people and puts their body parts in coolers and pretends to go fishing."

"They say Indians are coming here, back in the hills across the road there, and leaving things," I told them.

"What do you mean?" Tracy said. "Like body parts and stuff?"

"No, just things—sticks, brush, stones, pine cones."

"Why would they want to do that?" they asked.

"They say it is their way of honoring the place. They're trying to get back to their old religion."

"So they leave like sticks around?" Tracy asked.

"Yes, sticks. I've found pine cones placed in the crotches of trees, things like that."

"You mean like stick sticks?" she asked.

"Yes."

"Talk about weird," she said.

There actually was a dark event at this spot back in the 1920s, but given the discussion of body parts, and the slim, cloud-filtered, quarter moon, the half-naked younger generation, and the presence of a man such as myself, who, I am sure, would be the subject of much conversation and teenage folklore for many months to come, I decided not to tell them.

On such a night, in August in the late 1920s, a local swain, who could not have been much more than a teenager himself, was rowing his girlfriend on the dark waters of Nagog. Somewhere, somehow, between the island and the western shore, where the teenagers and I were now standing, this woman, Virginia Mills, fell, or was pushed, from the boat. She had a flashlight in her hand at the time, and as she sank beneath the water, her companion, or paramour, or whoever he was, saw the light sinking, an eerie gleam, descending deeper and deeper below the surface. Later that night the rescuers saw the faint glow and found her body.

The whole thing was reminiscent of Theodore Dreiser's *An American Tragedy*, which was published in 1925 and would have been known at the time of this incident. One wonders whether it was a question of life imitating art.

The presence of uncouth teenagers, bathing and, for all I know, relieving themselves in the waters of the good people of Concord, made me think I should, after all, do something for the community, so I told the teenagers the story of Sarah Doublet and the horned water monster that the Indians believed lurked in the deeper parts of the pond.

"It had these long tentacles, they say, and a huge gnashing beak and horns on its head. At night it would reach up and feel along the shore for people, fishermen, swimmers, things like that. If it caught you, it would either drag you down into the waters, or worse, slice you open and suck out your intestines."

The teenagers were quiet for a minute.

"You mean it would like eat you alive?" Tracy asked.

"Yeah, suck out your inner body parts while you clung to a tree."

"Cool," she said.

A week or so after that incident, I was once again up on the wall in back of Rick Roth's pig yard when I saw a stocky man proceeding through the trees, stopping periodically to mop his brow and look around. He was wearing a porkpie hat, heavy dark pants, a white shirt, and a windbreaker. He seemed to be looking for something, and when he spotted me, he came straight for me.

"Seen Mr. Morrissee?" he asked after the requisite greetings.

He had come out from Boston, he said, to look for Mr. Morrisee. He was a stubby man, half-shaven, with a fine gold front tooth and hands that were so thick and callused they seemed to be made of wood. He spoke with a thick Italian accent. I chatted with him briefly and then, in Italian, I asked why he wanted to see this Morrisee fellow.

"Was a good man," he said in English. "Only boss I ever worked for who would take care of you. Only boss to pay what he say he will pay when he say he will pay. No union man. No papers. No contract. He take care of you. I'm come here to thank him. Me, I'm retired now, but I never forget Mr. Morrisee."

I invited him to sit with me on the wall for a spell, but even though he demurred, he stayed talking on his feet for a long time. I told him what I knew about Morrison's current operation, and he told me about the construction jobs Morrison had been involved with–Fort Devens in the nearby town of Ayer, Columbia Point in Boston, and several other large works.

He couldn't get over the fact that Morrison was raising apples in his supposed retirement.

"He could get a big condo in Florida if he wanted. He's got gold rings. Got a lot of money."

He wanted to know who did all the work on the orchards, so I told him about the Jamaicans Morrison imports to his property each season. Dennis had recently told me a story involving a heavy engine block that he and his partners had to lift onto a truck bed. The narrative was hard to follow, but apparently Dennis spotted a jack nearby and proceeded to use it to hoist the engine, which he would then shore up with wooden timbers and then hoist more. It was a slow way but very practical. The old man came by, saw what they were doing, and made them take it all down and lift the machine with brute strength, not because it was the easiest or the most practical way of lifting the block, but because it was the way he had told them to do it in the first place.

I related this story, as far as I understood it, to the Italian man. He shoved his hat back on his head, stared off at the distant pond, and smiled, revealing his golden tooth.

"That's Mr. Morrissee, all right. But he pays good. No unions."

My newfound friend announced that he should be off to look for his former boss, but I managed to ask him where he was from in Italy before he left—I had recently heard of a valley in the Cadore district where an ancient pre-Christian system of distribution of common lands endures to this day, and there were said to be others scattered around in isolated sections. My friend was from the mountains of the Abruzzi, so I asked if he had ever heard of the type of valley or land distribution system that I had heard of.

"Not there," he said. "Not any more. Old times they had that. But things have changed. Now we're all just plain poor, come to America." He winked at this, and tipping his hat, bid me farewell.

It is unlikely that the Christian people at Nashobah understood entirely the concept of the fee simple, of absolute ownership of property, but by the 1670s they must have been getting the idea, and they certainly understood that the English would give them material goods in exchange for what the Indians understood to be the use of the land they occupied. The people living in the Praying Indian villages also understood that in order to make this exchange they had to petition the General Court. And once the permission to sell was granted, they sold. After the war, and the breakup of Nashobah, many of the Indians who once lived there and, seemingly, a few such as John Speen and Thomas Waban, who merely had family there, were trying to get permission to sell, although there were many who attempted to hold on to what became known as the homelands. The whites who had illegally carved out estates within the Indian lands were also trying to get title to the lands, but the General Court was careful about these transactions. For example, one white, Joseph Wheeler, a highly respected citizen of Concord, petitioned to buy Indian land at Nashobah and was denied by the court.

I am not trying to argue here that the English were in any way benign in their treatment of the Indians. But given the zeitgeist of the seventeenth century in general, the Puritan fathers, whom we so love to hate, did exhibit a certain fair-mindedness. They were attempting to Christianize and civilize what they saw as a lost culture, one of the ten lost tribes of Israel, as John Eliot and some others thought the Indians to be, and one of the most important aspects of European civilization was ownership of land. It was a means of becoming human. If you didn't own land, you couldn't vote. This is a fact that is still with us, as homeless people with no fixed address know all too well.

There are not a few critics of Eliot around today who prefer to use the phrase "internment camp" for Eliot's Christian Indian villages. It is certainly true that the plan was to have all the Indians living in villages, where they could be controlled and watched, and where the land would be cultivated in an unchaotic fashion and redistributed in an orderly fashion by purchase and sale. And it is also true that, after the war broke out, there were among the English a certain element who would have been happy to exterminate all the Indians. This was especially the case among the "outlivers." It was these people, who had themselves been most vulnerable to raids, who would have been responsible for the attack in which Sarah's son was killed.

But the fact is that the Puritans were of a different stock than those who came later to this country and also different from those who were settling in the South and those who later settled the American West and the Canadian North. The Puritans were a moral people who struggled with ethical issues, and they did not act in a rash manner. The argument for the praying villages, and later for the outright internment on Deer Island of all the Christian Indians, was that they had to be protected from the radical elements within society who would have killed them simply because they were Indians. Furthermore, the missionary work of John Eliot and Daniel Gookin, Thomas Mayhew, and Roger Williams sprang from a deep spiritual belief in Christianity; it was not some nefarious, racist extermination plot as has been charged in some quarters. The great seal of the Massachusetts Bay Colony sums it up. The emblem depicts an Indian waving a cornstalk, declaiming "Come over and help us," and the original charter for the colony mandated that the settlers "win the Indians, natives of the country, to the knowledge and obedience of the only true God and Savior of mankind, and the Christian faith."

Nevertheless, whatever the mission, the land was changed by the English traditions. The natural boundaries were legally abolished, and the land was distributed according to a mathematical formula, a measure of miles, a geometric form, four miles to a side, laid on top of the land of Nashobah with little regard for the river valleys that once defined the people as a tribe or band, or, as some believe, even a linguistic group. What remains, in our time, is the straight lines of the surveyors, the angles, elevation lines, a human construct laid over a natural landscape.

One hundred and ten years after the closing of the Nashobah village, the model was laid out across the whole continent. Within another century the Europeans spread over the whole territory, trailing fences and new boundary lines behind them with never a thought that this might be a wild land with an ethic of its own that would not necessarily allow itself to be shaped into the little boxes that Western civilization chose to lay out. Until fairly recently, when the embryonic legal rights of living things and a more holistic view of nature have begun to form, grandiose plans for the mastery of nature were considered signs of progress, a hopeful, bright future in which man (literally in this paradigm) would endure.

The monuments of this attitude are spread now across the American West like the temples, stellae, and pyramids of some ancient lost religion—immense concrete dams that block the free flow of rivers, tracks of railroads and ribbons of concrete and asphalt bisecting a whole continent, pylons of electric wires webbing the wildest desert places, cities so squared and paved that they have obliterated all but the tiniest plots of greenery. The grand designs are not over yet in spite of their obvious drawbacks. As recently as the 1960s, engineers were eyeing the waters of Canada to irrigate the deserts of the American West, and even today Canadian engineers construct immense lakes nearly the size of France at James Bay to create electricity.

How did it become legal to tamper so freely with nature? How did it come to pass that cultures such as those of medieval France and England, which used to put wild animals such as rats and mice on trial for raiding 157 granaries, could evolve into nations that can actually own outright the natural world, not only land, but rocks, plants, and animals?

In American law, according to the Solicitor, the first case appeared in 1805, the same year that Thomas Jefferson entered his second term and laid the groundwork to have the United States surveyed.

In May of that year, a certain Mr. Lodovick Post set out with his hounds and horse in pursuit of what the law books refer to as a beast, *fera natura*, which is to say, a wild animal. The beast in this case was a red fox, an animal despised by Mr. Post, the plaintiff in this particular case, and also by the defendant, a certain Mr. Pierson. In fact, at this point in time, foxes were despised by almost everyone, save a few eccentric nature lovers like John James Audubon, who was then hard at work on his *Birds of America*. Accordingly, the law, and in fact the decision in this case, which ended up as a precedent for American property rights, offered the greatest possible encouragement to destroy this cunning vandal.

But all that was merely incidental. What counted in this case, the Solicitor says, was ownership of wildness.

Mr. Lodovick Post, a country gentleman of some means, had mounted his steed at the crack of dawn on the day in question and with hounds and horn rode out in pursuit of the "wily quadruped"–as one of the lawyers in the case called the fox.

"Toward evening," the Solicitor says, "having pursued the windings of Sir Fox the day long, and weary with hunting, Squire Post closed in on his prey, whereupon Mr. Pierson–who had not shared in the labors or the honors of the hunt–appeared on the scene. This Pierson drew his fusil, took aim, shot the fox, and made off with it.

"Who owns the fox?" the Solicitor asked. "Mr. Post, who chased it all day, who tore through brambles, who sweated, who went without lunch, who was stiff with riding, whose hounds were hoarse, breathless, and scratched? Or Mr. Pierson, who happened upon a tired fox and shot it?

"Post took Pierson to court, claiming ownership, and won his case. Pierson appealed, and there rose a mighty furor over the poor, long-dead fox, who by this time was reduced to a mere pelt.

"Counsels for both parties reached deep into history to make their points, citing, among others, the Roman emperor Justinian, whose laws held that pursuit alone vests no property. Pierson's counsel argued that the fact that Post rode after the fox all day meant nothing, mere pursuit gave no legal right to ownership, and the fox in fact became the possession of Pierson as soon as he killed it. Post's counsel countered that whatever Justinian thought about foxes was insignificant. He argued that these were modern times, this was 1805, mind you, and society had changed since Roman times, and if society changes, should not laws also change?

"Post lost the case, Pierson got his fox after all, and *Pierson v. Post* went on to become a seminal case in the annals of American property law."

But the dissent bears watching, according to the Solicitor. In modern times, in a world in which "resources" (a loaded, anthropocentric, word if ever there was one) are shrinking, who in fact does own the fox?

"Sir Fox," the Solicitor says, "were he with us now, would surely argue that he is his own man and beholden to nobody but his wife and little ones."

The Indian people of the Nashobah area would have had a problem comprehending this case, since a man returning from a hunt, no matter how long or laborious it may have been, was bound by tradition to *share* his catch with the other members of his tribe.

But if the ownership of a fox or any other quarry is in question, it's not much of a jump to wonder who owns the land? Once you *have* a thing, it's yours, according to the Solicitor. Outright possession, as in the case of the fox, is a statement of ownership.

"It's a way of communicating to the world that this thing is yours, fox, land, mountainside, doesn't matter. Property law, as it is currently understood, is therefore"—he's winking here and adopting his legalese tone of voice—"is therefore nothing more than an expression of control over the natural world. *Fera natura,* my good man, is no more *fera* than it is *natura.* Not in the legal world of affairs."

The Solicitor pauses at this point and takes a healthy draught of his Guinness. He replaces the glass on the table and lifts his index finger with Dickensian flair. (I have heard that this man is a skilled courtroom dramatist, although I have never seen him in action.)

"There are, how shall I put it, *Extenuating Circumstances.* That is to say, complicating factors. Put the case that A owns a mountainside and B owns the land below the mountain. Heavy rains for thirty days and thirty nights. Thunder rattling in the dry hillsides. Soils soaking and muddy, and then one night a great rumbling, a vast roar that fills the air around the mountainside. B comes out the following morning and finds that the whole side of A's mountain has slipped down over his property, covering it to a depth of some ten to twenty feet in its entirety. A appears on the scene, surveys the prospect, and says, in so many words, Ah ha. All this is now my land.

"Who indeed does own that land: A, whose mountainside it was, or B, whose land is buried beneath A's topsoil?"

The glib Solicitor later admitted to me that he stole the case from a fictional story by Mark Twain. But there were similar cases in history, he says, one as recently as 1973, when Omaha Indians along the Missouri River sued to recover land that had been lost to them in the late 1800s when the river changed course and cut away several thousand acres of reservation land on the Nebraska side of the river, where they lived. White settlers soon moved in and staked claims, and the Indians, who barely understood the concept of private ownership of earth, paid no attention. But by the great awakening of the 1960s, Indians understood all too well the legal ramifications of private property and were using the courts of American law, in case after case, in an attempt to recover land that was stripped from them by those same courts, under questionable legal practices.

In 1973, the Omaha, led by a standing member of the Omaha Tribal Council, cut a fence blocking their ancient property and occupied the land, claiming ownership and citing the original 1854 treaty granting the territory to the tribe. Owners charged trespass. The case went to a local court, and then slowly worked its way up through higher and higher courts until finally, after a few initial victories in which the Indians gained, or regained, part of the territory, the case began shifting around in the federal courts. In the end, the Indians got some of their land, and the whites held onto some of what they considered their land.

But one of the unfortunate things—or fortunate, depending on your point of view—is the fact that Indians all over the United States, having learned the value of private property and the benefits of the legal weapons of conquering European cultures, are transforming themselves into masters of Western materialism. There is now a casino on the regained Omaha lands.

The Omaha had to be taught to divide their lands into lots. In 1881, not long after the massacre at Little Bighorn, an enterprising woman from Boston named Alice Fisher set out to live among the Omaha, ostensibly to learn about their culture. She ended up teaching them about her own. One of the things that disturbed her was the fact that the Indians did not seem to have a concept of holding land as property; they shared everything. She began working with government officials to have an act passed that would grant each Indian head of household 160 acres in trust for twenty-five years, after which, if they had learned their lessons, the Indians would own the properties outright. The act passed in 1882, and by the early part of the twentieth century the Omaha were granted legal title to "their" lands.

The first thing the Indians did was to sell. They needed money to pay debts they had accrued. Land sharks moved in, and within a few years the Indians had lost all their lands.

One of the key moments in Indian history, as legal authorities on the subject of Indian lands know all too well, occurred in 1887 with the passage of the Dawes Act, or the General Allotment Act. It was, in some sense, the end result in American law of that which was begun with English law and the work of John Eliot to allow the Indians to own private property at Nashobah with accompanying rights of sale. The Dawes Act had the support of both Indian sympathizers and land-hungry profiteers, and was not seriously contested. The concept was that all Indian lands would be divided into privately held, single-family plots that could be bought and sold at will, regardless of the desires of the tribe as a whole. Indian historians now see it as a double-edged cut: on the one hand, it broke up the land, but worse, it broke up the idea of communal land, of council and consensus, and transferred it into the ideal of private ownership. In effect it broke the traditional culture of the Indians. This was the point, of course. Washington viewed the establishment of private property among the tribes as the primary civilizing force. "The common field is the seat of barbarism," one contemporary bureaucrat said.

It was a profoundly American statement—never mind that the civilizations of the Middle East, ancient Greece, the Han Dynasties, Rome, Meso America, and the city-states of the European Renaissance and medieval Britain all made use of common fields—here in America, civilization was private property.

Shortly after passage of the act, surveyors set out for the Indian territories and began dividing the lands into neat packages of 160 acres per family, which was the amount deemed necessary for survival, providing the land was well watered, of course, and the soils fertile. Then census takers tallied the number of Indian heads of household within a given tribe and "gave" them their allotments. After the lands were assigned, the remainder was put on the auction block and sold to the highest bidder. This was in effect the plow that broke the plains.

It was by these stratagems, some well intentioned, it must be said, that the Indians lost North America. In the 1850s Indians still held almost the entire western half of the continent. By the end of the 1860s and the beginning of the Indian Wars, the whites had chopped into their holdings from both east and west. By the turn of the century, except for the Southwest and a small section along the Canadian border, it was all little pockets of reservation land.

None of this was confined to North America, of course. In Mexico and Central and South America, the same thing happened. Spanish settlers rooted out the commons and the rights of use, passed down in most cultures through the female line, and replaced them with a system of privately held tracts passed on through the male line. In Mexico, in areas such as Chiapas, the peasants managed to hold on to the small, communally held farms known as *ejidos,* where they grew traditional native crops. But the commercial agriculture of the great *haciendas* run by the Spanish, using peons as laborers, altered the traditional watering patterns and destroyed the native crops.

The fight for common lands is still going on in Mexico. As recently as 1995 the peasants of Chiapas attempted to reclaim or reestablish the system of the *ejido.* The rebels looked back to the Revolution of 1911 for inspiration and the peasant messiah Emiliano Zapata, who did not believe in the private ownership of land and fought to break up the *haciendas.*

These contemporary peasant revolts in Mexico are the exact opposite of current land-use "rebellions" in North America. In the past ten or fifteen years, groups of private landholders in the western United States have banded together to create organizations whose primary purpose is to restore, regain, reclaim—the phrase varies from group to group—the public lands of the West, which in this case happen to be held by the federal government, as National Park, National Forest and Monuments, or Bureau of Land Management properties.

These groups, collectively know by the euphemism "wise use" groups, have resounding names: Our Land Society, National Wetlands Coalition, Western States Public Lands Coalition, and People for the West. They sound very much like environmental groups, but in fact they are singularly laissez faire in their attitudes toward the use of land and would willingly open up ecologically sensitive wilderness tracts to mining, grazing, and timber cutting. Land, in the view of the members of these groups, is a resource, not a spiritual entity. You get your living from land, and God, in his wisdom, gave man dominion over all living things to do his work on earth—or so it is argued.

Membership in these groups is large, and the pitch is that this is a people's movement, since it has a large following. But in fact the groups have the backing of some of the largest corporations in America, and some powerful nonprofit national groups as well, such as the National Rifle Association. People for the West! is backed by the mining industry, for example. The National Wetlands Coalition, which sounds for all the world like a wetlands protection group, is backed by land developers, agribusiness, and oil and gas interests. These groups also have the support of a number of U.S. senators and members of Congress, some of whom have characterized the Environmental Protection Agency as the Gestapo of the U.S. government. These men and women work to block wilderness bills and the Endangered Species Act (the bête noire of the property rights movement) and to open up National Wildlife Refuges to oil drilling and National Forests to clear cutting.

Like the peasants of Mexico, the members of the North American groups are sometimes violent in their desire to resist the government. One member of a group told a friend of mine from Utah that he actually sympathized with the bombing of the federal building in Oklahoma City. "They were on the right track," he said.

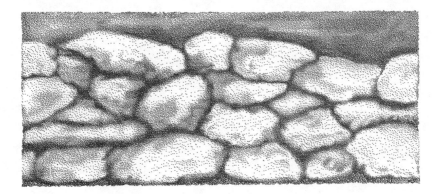

Chapter Twelve

Others now living or recently deceased remembered Indians, or had seen their ovens and traces of huts near the pond, on the "island" as we call the hard land surrounded by water and swamp, and on or near the flat ledge where the mountain cranberry grows, but it must be remembered that these later Indians were squatters, who came and went by sufferance, and I believe I can suggest a good reason why they camped on the island near the pond. Their ancient burying-ground was there. . . .

–H. J. Harwood, in Proceedings of the
Littleton Historical Society, 1896

Islands of the Dead

As a matter of curiosity, since I live near Beaver Brook, I attended a few of the meetings organized by the opposition to the Beaver Brook project, to follow the course of the strategies of the people fighting the development. The meetings were always held in the house of a man who, in his middle years, had decided to take the cloth. There were crucifixes and portraits of the Virgin all over his house. The land they were fighting to save was, as I said, probably just outside the boundaries of the sixteen-square-mile tract that made up the original Nashobah Plantation, and the weir where Sarah's husband Tom spent his last days was just upstream from an area that, if the developer got his way, would be someone's back lawn. At one point during one of these neighborhood meetings, I mentioned that this was Indian land once and that there was a remote possibility, very remote I said, that the deed of the five-hundred-acre tract could be questioned. That would perhaps throw the whole Nashobah Plantation issue up to review and slow down or scare the developer.

"Very long shot," I said.

The religious man reacted strongly.

"Let's not get into that," he said. "Let's not even touch that. I don't want a big casino in my backyard."

He had a point. In the end, it's not who owns the land, or whether it's common land or private land, it's a matter of how whoever controls it takes care of it.

One of my brothers lives on the coast of Connecticut, not far from the site of the highly successful Foxwood Casino. For years, on my way to visit him, I used to drive through the area on pleasant country roads through what I considered one of the finest landscapes in the East. The region evoked all that we seek in the Western version of the idealized landscape—small dairy farms, rolling hayfields, distant wooded hills with streams. In those years, one of the few eyesores in the area was a large ugly billboard on the tribal lands that advertised the Pequot Indian bingo parlor. But, after a prolonged battle for tribal recognition and a subsequent battle with the state of Connecticut to permit gambling on the reservation, the Mashantucket Pequots had constructed a veritable palace in the forest.

My brother, who is an archenemy of developers and, admittedly, biased, kept me informed on the construction progress of the new casino.

"First thing they do," he told me, "the Indians go in there and cut down a grove of ancient hemlocks. Beautiful feathery trees with a dark, mysterious interior. Stripped earth, big boulders lying around exposed to the sun. What kind of Indians are these? All this Mother Earth and Great Spirit talk."

Month by month, the landscape of the reservation began to change. Tall, urban buildings began rising out of the surrounding woods, visible for miles around. Wide parking lots appeared, raw earth, with yellow bulldozers running to and fro like worker ants. A few months after the Foxwood Casino opened, it began making money. Within a few years, it was reported to be the most successful casino operating in America. Huge busloads of visitors, many cars, much traffic, and so much business that the casino had to maintain a constant police vigil to direct traffic and visitors. Soon, the development began to expand outward. The now-rich Indians began buying up land around the casino; they put up a new hotel and opened new restaurants. In essence they began to create, in the middle of rural eastern Connecticut, in an area the tourist bureaus used to term "the quiet corner," a miniature version of Las Vegas.

Some of the local non-native people liked the project. If nothing else, it added to their property values. They could sell to the Indians at inflated prices. But many, including, of course, my brother, and even a few traditional Indians, hated it. One day, driving past the front gate of the Foxwood enclave, my brother spotted what he believed to be the ultimate metaphor for the whole venture: a red fox, dead on the road, its furry tail blowing in the wind of the indifferent passing cars.

Even though he has never been inside the place, he told me many stories of life on the inside—men so taken by their winnings at the slots that they pee in their pants rather than desert their machines. Children left locked in cars in hot parking lots while parents gamble away their savings.

"One day at the slots, a poor man had a heart attack," my brother told me. "Died at his machine and lay there on the floor. Surrounding gamblers glanced over and continued playing. Medics came and hauled him away while the drones at the slots carried on, like zebras who go on grazing after the lions have downed a fellow member of the herd."

More of an issue, and one that the officials at Foxwood seem to recognize, is the fact that people have gambling problems. My brother claims that there are loan sharks and lawyers waiting in the wings of the casino to prepare documents for those unfortunates who, having gambled away their life savings, want to mortgage their properties to get the money for one more chance at the big win.

I heard another story, albeit twice removed, of a man slumped over the hood of his car, crying his heart out, having lost, in a single afternoon, his entire life savings. I heard stories of suicides, one involving a woman who threw herself from a bridge after she gambled her life savings, then mortgaged her house, and then lost all to the casino.

Foxwood maintains a hotline for such people to get help and has literature stacked around the floors, but of course one of the symptoms of the problem, as with many addictions, is a failure to recognize that there is a problem in the first place.

Having heard all these stories, I couldn't resist verifying them, and so one winter afternoon, on the way to my brother's, I decided to visit the dreaded place myself.

This was one of those profoundly gray winter days when the seamless pall of winter hangs over the landscape and the snow fields are gray and the roads are gray and the white houses of rural New England are gray. If one has to go into a place where there is reportedly no natural light whatsoever, no view of the outdoors, it should be on such a day, I reasoned.

The drive down was uneventful. The road leading past the casino had been widened, there was more traffic, one or two of the fine dairy farms had gone out of business, and there were signs everywhere directing traffic to the place. Not that the signs were necessary. You could hardly get lost once you were within a few miles, since you could see the buildings towering over the surrounding trees.

Police were everywhere as soon as I got nearby: angry police for some reason, waving you this way or that, stopping cars. Litter in the parking lot—air freshener canisters, beer cans, of course, cigarette packs blowing in the winter wind—I was ready for that sort of thing. What I was not prepared for was the noise.

As soon as I entered, there was a grim-faced man standing inside the door who glared at me. I asked him if I was in the right place—he seemed to be posted there to keep people out, not welcome them.

"Right place for what?" he wanted to know.

The racket inside was almost deafening, loud piped-in music, bells going off, the clack and clang of the slot machines. No one was having any fun. Dull-eyed souls stood solemnly in front of row upon row of slot machines, holding large plastic cups filled with quarters, dropping them into the machines and methodically pulling the handles. It reminded me of industrial workers on an assembly line. Serious, tight-lipped men and women surrounded gaming tables, staring blankly at roulette wheels or card sharks. Leggy young women in high-cut green tunics and black net stockings, low-cut halters, and strange little Robin Hood—style hats with green feathers cocked at a rakish angle circulated among the players with trays of drinks, which, as long as you were playing, were free, I was told. No one was happy, not even the winners. I watched one man collect a huge number of hundred-dollar bills from a cashier's window.

"Did you just win that?" I asked. He didn't seem pleased, if he had.

"Yeah. Now I'm going to go lose it," he said.

The place was actually like a small roofed-over town devoted entirely to entertainment and shopping. There were movie theaters, restaurants, bars with performers singing into microphones, and loud bands blaring, even now, early on a weekday afternoon in the middle of winter when, according to my worldview, everyone should be working at something or sitting home by a cheery fire, with a pot of tea, a good book, and the cat purring at the hearth. This was canned happiness, canned life, bright colors everywhere, screaming music from hidden speakers competing for dominance with the live band. The rough carpeting which covered everything but the bathrooms had a garish red swirling pattern that was so ugly, so hideously disturbing, that it actually made me feel sick to look down.

But I couldn't look up either. Ugly there too—bright lights, a bizarre, twisted ceiling pattern. And I couldn't look straight ahead. Grim, sour people, more lights, ugly machines. I thought I better leave, I'd had enough already, and then suddenly in front of me, one of these guards again.

"What are you doing here?" he wanted to know.

I was so taken aback I could hardly answer.

To record the ambiance and get some quotes from the happy gamblers, I had brought a small tape recorder. The guard was not happy about it.

"What's that thing?" he wanted to know. "You can't record here."

I tried to explain my mission, which I admit was not easy: What was I to say? I'm investigating the fate of lands held by Indians? The evolution of the concept of private property out of medieval English common law? The conflict between privately held land and land held in common? Proper stewardship? Whatever I mumbled in my confused state served only to increase his suspicion, and after some more interrogation I was instructed to go directly to the public relations desk and surrender my recorder, whereupon I would be given a packet of information about the Foxwood Casino. Seemed like a good escape, so I walked off directly. I could feel his eyes on me as I fled.

At the public relations desk an indifferent woman filing her nails said she didn't know anything and gave me two brochures and a number to call, which I did. A nice man said he would send me a packet of material—which he never did. No one asked me to surrender my recorder, so I continued to wander around some more; in fact, by now I was lost. Everything began to look the same, but down in the basement, more or less hidden from the masses, I found the "museum" my brother had told me about.

"They've got a tepee there—Western Sioux," he told me. "They have a moth-eaten stuffed red fox, which some biology type told me wasn't even a local species back when the real Pequots lived here."

There was indeed a stuffed red fox there. And there were one or two visitors wandering around, looking at Indian artifacts, a few tools, and arrowheads, and then at the back of the room, an audiovisual account of the 1637 Pequot massacre.

The Pequot Indians had once been a powerful, warlike group who had settled in coastal Connecticut shortly before the English came on the scene. Because they were independent and feisty, they soon got into disagreements with the invading English over property and trading rights, and put up a resistance. The English determined to put an end to the struggle, and after a few initial skirmishes, surrounded the Indian fort at present-day Mystic. The English were led by Captain John Mason and the soldiers were accompanied by Mohegan Indian allies. Using local Indians to gain information, Mason decided to eliminate entirely the Pequots in the village.

At dawn on May 26, 1637, the English opened fire on the stockaded town and then set fire to it. The Pequots, taken by surprise, put up a resistance, but they were armed only with bows and arrows, and as they ran out from the conflagration, the soldiers picked them off. It was the equivalent of a turkey shoot. Some five to seven hundred men, women, and children were either shot or burned to death, "roasted" as one contemporary wag accounted. The English lost all of two soldiers: in short, a brutal massacre.

I got the point of this whole Foxwood Casino operation right away: revenge.

Out of the thousands of people I saw that afternoon, none appeared to be Native American save perhaps one or two of the attendants working there. Thousands of white people, day after day, week after week, were donating their money to the Mashantucket Pequot. It was their reparation for the damage done that May morning at Mystic in 1637.

Revenge, perhaps, but on whom? The poor fox dead on the road, all that uprooted land, and all the poor, addicted souls dispensing their life savings? And what about the fact that, as far as I could see, none of the people giving donations to the Pequots had much money to begin with, nor were they or their ancestors at the massacre.

In the museum, along with some historic literature, I found a couple of brochures about the casino. One of them was a guide to the games: blackjack, craps, baccarat, Big Wheel, and something called Omaha Hold 'Em, which had nothing to do, as far as I could tell, with the Omaha Indians who fought so hard to regain their land along the Missouri River.

The other document described the "resort," as they called it. I was invited to "experience" gaming in its "natural state"–whatever that means. I was instructed to "choose" Italian, Chinese, and American delicacies from their dining rooms and to find all the charm of a "quaint country inn" nestled in the woods "200 feet" from Foxwood Resort, which, unless that was a misprint, would mean that the country inn was located in the parking lot for the gambling rooms. There was shopping and a city of specialty theaters, Turbo Riders (whatever they are), arcades, virtual adventures, and all the other pleasures and entertainments of the American dream. The brochure was illustrated with glossy photos of happy people gaming–happy white people, by the way–shopping, winning, riding on machines, and dining. In each of the photos, I noticed, the people were smiling broadly. It was a sharp contrast to the stony-faced players out on the floors.

Mashantucket, I was informed by the brochure, meant much wooded land, and the Pequots were known as the Fox People. But beyond the casino walls, the woods were stripped bare, and out on the road, the fox lay dead.

I left the place and drove on to my brother's. It was all so sad. To my knowledge none of the local Pequots came out to stuff a few tobacco leaves in the fox's mouth for its spirit journey to the next world. No one stroked its nose three times before sending it off. No one propitiated the fallen hemlocks and the tree spirits. We're all just good Americans now, Indians and whites alike, and this land is our land to do with as we please because we own it in fee simple and are no longer beholden to anybody, neither the lord of the manor nor the lord of the universe.

How, one wonders, did it all come to this? More to the point, perhaps, how did the Pequots, who were nearly exterminated in 1637, maintain themselves as a people in the face of disease, Christianity, and the all-powerful forces of the English and, later, the American culture?

One of the fortunate outcomes of the new casino is that the Pequots are now building a new museum and research center to help explain their history. They have also funded new archaeological research in the area and have recently turned up evidence of an Indian fort, along with artifacts. The new thinking is that the people endured as a tribe long after the supposed massacre.

Town histories report that as late as the nineteenth century Indians endured in the Nashobah area as well. They used to reappear at the site of the former village and camp for a few weeks on what were known as the "islands." These islands were never identified properly; they may have been the actual islands in Nagog Pond or spots of high ground in the wetlands surrounding Nashobah. In either case, the story was that these camping areas were the gravesites of the ancestors of the people who had once lived there. The Indians would appear mysteriously, stay for a week or so, and then disappear. No one knew who they were or what part of the country they had come from. There are similar stories of Indians reappearing at certain sites around New England in the nineteenth century. One indefatigable journal keeper of the early nineteenth century wrote that a whole town near Deerfield would turn out to watch the Indians who would appear each year at a bend in the Connecticut River to camp for a few weeks.

I had heard from one local source that the island that lies just off the Morrison property and belongs now to the Concord Water Department was an important site for the Indians of Nashobah. The woman who told me, a member of the Acton Historical Society, had visited the island and said that there were a couple of stones out there that looked like grave markers and that they were arranged in such a way as to match, she thought, the rising sun at the autumnal equinox.

This was intriguing, since I had been to the Praying Indian burial ground at Christiantown on Martha's Vineyard, where a separate and more successful Christian Indian mission work under Thomas Mayhew was carried out at the same time as Eliot's experiment. Mayhew, who was one of the earliest white settlers on the island, had the eye of an anthropologist and a good ear for the Algonquian language (unlike Eliot, so I have read, whose sermons may have been only half comprehended by his budding parishioners). Mayhew converted the sachem of the island, and then this same man began converting others. There was a thriving Praying Indian village on the Vineyard long after the others on the mainland had failed.

The burial ground for these people is a sad little clearing, now maintained by the local Gay Head Indians. Save that they are arranged—more or less—as the headstones of a typical New England graveyard would be, you would not necessarily recognize the burying ground; it is simply a collection of granite boulders propped on end. The description of the stones on the Nagog Pond island sounded similar, so I decided, in spite of the fact that the island is off limits, to make a voyage of discovery.

When to undertake the expedition, though? Cover of night would be the logical time, but then I wouldn't be able to see the stones. Weekdays the road along the pond is often busy with runners and local walkers, not to mention the watchful eye of the local police, who cruise by there periodically in search of renegade fishermen and swimmers. I finally decided to choose a Sunday morning, the quietest daylight period. So one still autumnal morning, I parked my car away from the pond and portaged my canoe through the wooded thickets to a likely launching site on the peninsula where I had discovered the swimming teenagers.

The waters of Nagog, unlike the waters of some of the other ponds in this region, are very clear, and I could see ominous caverns and hidden worlds in the brown gloom beneath the surface. Were this not a bright Christian holiday and also the secular twentieth century, I should have worried that the slimy tentacles of the horned water beast would slither up over the gunnels, grab me by the neck, and haul me under to its lair. I did see at one point a slithery motion in this same area on another expedition but determined it to be an eel.

Now there were only a few delicate, white and gray seabird feathers on the surface and a few red-splashed leaves from the nearby maple trees. No cars on the road. No people.

There is a sheltered cove on the south side of the peninsula, hidden from all passersby, and before making my dash across the open water to the island, I poked around in the shallow waters, looking for turtles.

One of the more attractive elements of Nashobah is the presence of these local ponds, which is what brought the Indians to the area in the first place. These "lakes," as they are called locally, are small and, except for one in the nearby town of Westford, do not attract the plague of powerboats and jet skis that destroy the peace and quiet of so many lakes in this fast-paced United States. The local town lake, whose southeastern bank Linda Cantillon was fighting so hard to preserve, had a limit on the horsepower of outboards, and this simple regulation had the effect of keeping all but fishermen and canoeists away. Those who appreciate high-speed, powerful engines do not appreciate regulations of any sort, it seems to me. One restriction—one minimal curtailment of their "rights" as they call them—and they seek a freer environment. It is the same mentality, I think, that drove the American settlements westward to the lands of the free.

Nagog of course is off limits altogether, so I was alone that day except for the crowds of herring and black-backed gulls that collect around the rocks on the south side of the island. This was late September, and out on the open water beyond my cove I could see little flocks of scaup, dipping and diving, and perched on the rocks around the island were two or three cormorants drying their wings—an omen I thought. The Puritan poet John Milton associated the cormorant with the archdemon, and they appear, along with owls, in the dark prophecies of Isaiah. Having long since deserted my Puritan roots, I felt no fear, but they did lend a fitting cast to the island, given the fact that this is the supposed burial site of the Indian sinners of the seventeenth century, as well as the site of the darker events that transpired on that summer night in the 1920s when poor Virginia Mills sank to her watery grave.

Even before this first visit, I had begun to equate this otherwise pleasant island with the traditional Isle of the Dead, where, according to the Greeks, the souls of warriors go after they have died. There is a painting of the isle by the nineteenth-century German Romantic Arnold Böcklin, a haunting image of a high island with dark cypress trees, which for some reason has stuck in my mind for years and which I had begun to associate—not without reason—with the mystery of the dark-treed island at Nagog. The white pines, which stand like sentinels on the slopes, even on bright autumnal mornings, lend a dark and foreboding aspect to the place. But, summoning my courage, I made for the island anyway.

Quietly, I paddled up to the last point of land on the peninsula, and then while all was clear on the landward side, I made a dash for the eastern bank. Kneeling amidships in the canoe and paddling furiously, I gained the hidden side of the island in a minute or two, presumably unseen by any shore-bound watchers; then approached from the east at a more leisurely pace.

The water on this side was shallow, the bottom rock strewn, and some of the flat boulders were only a few inches under the surface, so I had to skim over them or around them lightly. The banks of the island were thickly lined with buttonbush and sweet pepper bush, growing in such a dense tangle that I couldn't find a landing spot until I got to the far northern end of the island, where there was a little deeper water and a cut in the bushes. It turns out I was not the first person to have come ashore here since the Concord Water Department took over the property.

The first thing I saw when I gained the ground was a man standing in a grove of pine trees on the highest point of the island. On second look I realized that someone had draped a coat and pants from a high tree stump. It was a bizarre image, a little like a hex sign or a warning. I moved up the bank and came to an old foundation. Up until the 1940s, there had been a big summer house on the island, owned by a doctor from Newton who had a large extended family and who would move here in April and stay through October. Surrounding the remains of the chimney and the foundation were many scattered artifacts, presumably left by the good doctor and his family. I saw some tin basins, a cooking pot, the remains of a kerosene heater, and then, in the middle of the chimney foundation, I discovered the archaeological evidence of more recent occupation, a nest of beer cans. My marauding, night-dwelling teenagers no doubt, although how they got out to the island was a mystery. The cans turned out to be full.

The gravestones, I had been told, were on the south side of the small island, which is no more than an acre or two all told. I walked over to the other side looking for stones, but after a futile search gave up and sat down on a large flat rock on the southwest shore. From here I could get a pleasant view back to the wooded banks of the peninsula and the rising ground where the serpent mound is located and, above that, the rock outcroppings of the granite-strewn woods. Sarah Doublet may have come to this same site, for all I know. The Nashobah women must have made periodic forays to

the island to pick berries, or perhaps to bury their dead, and so it is possible that she too repaired to this spot, although I doubt that she sat down to enjoy the vistas.

Now, as it was in her time, the hill is a solid wall of trees. By winter, from this same spot I would be able to see the veritable bones of the hill, whereas now it was all reds and yellows, and the trees and shrubs were fused with various shades of green. To my left, on the eastern bank of the mainland, I could see cars passing on the Great Road and a flat, angular collection of buildings that contrasted sharply with the rounded nature of the landscape in these parts. Beyond those buildings and south southwest there is another tract of preserved open space whose slopes also contain a series of angled walls, as well as a chamber, which some people believe was sacred to the local Indians. On the winter solstice at dawn, the rays of the rising sun will pass through a chink in the eastern wall and make a half circle around a small quartz stone on the northern wall. This, it is believed by students of the subject, is one of the markers in the great system of stone monuments and relics that descend on Nashobah and radiate outward into the larger universe.

With thoughts of alignment in mind, I had brought a compass with me, and since the rock I was on seemed to be at the southwestern end of the island, somewhat in line with the winter sunrise on the day of the solstice, I placed the compass on the rock, lined up the degrees of the rising winter sun, and walked back across the island to see if there was perhaps some marker on the eastern end of the island. On the way, I spotted an interesting green-capped mushroom, probably a russula, which I stopped to inspect. Then I found a dead black-backed gull and separated one of its beautifully marked wing feathers, and having absorbed myself with this for a few minutes, moved on until I saw the relic of an old potsherd from the '20s, with handsome blue patterns, a piece of willowware, maybe, or even Dedhamware. I picked it up and meditated for a while on the nature of pottery. Then I started thinking about the fact that this actually was someone's dinner plate and that here, in this spot, on summer evenings, the good doctor and his family would gather, having consumed from this very plate, perhaps, boiled corn from the mainland fields, and chicken and mashed potatoes, washed down with frosty iced tea poured from a fluted glass pitcher. And then, in this stillness of the morning, I imagined the breakfasts they had here, eggs and pancakes and the smell of percolated coffee, and then the thunk of oarlocks in the flatiron rowboats, the hot smell of wicker, of old canoe canvas and varnished gunnels, and the sound of the lap of the fresh water on hot sunny rocks. I turned the sherd over and inspected the fine network of cracked lines and somehow from this remembered stories I had heard about the daughter of the doctor who lived in this spot, a fine, high-boned woman with a black mane of hair, who kept a white horse on the mainland in the Morrison barn and used to ride out over the pastures and through the woods, who had violet eyes and wore white shirts, and was always gentle and fun loving and kind to the little farm girl who lived in the Morrison house, and who, at age eighty-three, had told me all this in her cracked old voice.

Here, with the soft present-day lap of waters on the rocky shores and the shimmering light on the leaves, I drifted into reverie after reverie on the nature of time and place, and sat down with my back to a tree, facing the morning sun to think some more. As the sun warmed my face I fell into a drowsy sleep in which I could almost see this place as it was. Now it is all dark woods and somber light even in summer, especially in summer, in fact, when the leaves are out. The only open lands left in this area, other than the short sweep of meadow down to the shoreline from Morrison's house, is the high mowing on the north side of the property by Long Lake, which Linda Cantillon was trying to save. But it was not always so.

If you look at the nineteenth-century maps for this part of the town, you will see all the privately held farms marked with the names of the owners and the pastures and closes, meadows, walls, and what they called "mowings" carefully indicated. All of this was open land, a wide sweep from this island up to the crest of the hill where the winds were pent, west to the crest of the next hill where the lone cottage of an old cowherd who lived on top was perched, and thence north northeasterly, as the surveys phrase it, to the low ridge of the northern side of the pond—all sky and green fields and bobolinks and meadowlarks calling throughout the summer. This was a different world in that time, a rural idyll with a fixed purpose to existence, and with such a definitive commitment to permanence that the landholders felt no concern about having their names and places of residences fixed to an official printed town map, as if this family and this property would be forever linked to one another. Such was the devotion of those who held land. They should have known better, must have known better in fact, with all the westward shiftings, and the opening of the Erie Canal, and the migrations of 1846.

I'm not sure how long I stayed at that spot. But, in time, I roused myself and wandered down to my canoe, launched it, and still daydreaming, drifted over to the southwestern side of the island. I pulled out from the shelter of the trees far enough to get a view back to the mainland, and then, seeing no

one around, once again made a mad dash for the sheltering waters of the peninsula. Once there, I slowly made my way back to where I had pulled out. It was not until I had reloaded my canoe on the car that I realized I had left my compass on the flat rock on the southwest end of the island.

Chapter Thirteen

The first man who, having fenced in a piece of land, said, "This is mine," and found people naive enough to believe him, that man was the true founder of civil society.

–*Jean-Jacques Rousseau,* Discours sur l'Origine et le
Fondement de l'Inégalité parmi les Hommes, *1754*

Fences Defined
Fences four feet high, in good repair, constructed of rails, timber, boards, iron or stone, and brooks, rivers, ponds, creeks, ditches, and hedges, or other things which the fence viewers consider equivalent thereto, shall be deemed legal and sufficient fences.

–Annotated Laws of Massachusetts, *chapter 49, "Fences,
Fence Viewers, Pounds and Field Drivers"*

Who Really Owns
North America?

I saw Farmer Flagg at his stand selling pumpkins a few days after I visited the Isle of the Dead and stopped to talk to him. I told him, in so many words, that I had been trespassing on land held by the Concord Water Department and that the land had been taken out of private hands for the public good and that I really should turn myself in because, in the end, the regulations were promulgated to protect a public resource and that this, in the end, is what most land restrictions were all about. It was a setup, of course, and it got him talking. But then it doesn't take much to get Farmer Flagg talking.

"Isn't there some justice to these laws?" I asked. "Aren't they just trying to protect the greater good?"

"No," he said. "They are trying to protect the political office of the incumbents ... Seriously though, if they were going to build a school, wouldn't they just buy the property? If they were going to put a highway through this farm, they'd offer me a fair market price, buy it, and develop it. I don't have a problem with that. I've got a problem when they come here and say you've got a blue-spotted salamander in that pond back there. You can't develop this land. Fine, I'd say. I won't build on it. But pay me what I'm going to lose by not developing. Pay me the difference. Only fair, right?"

"I suppose," I said. "I'll have to ask my lawyer."

"Ask. He'll agree. I talked to a lady at the stand this week, she and her husband had bought a double lot with a house on one and the other undeveloped. They checked with the planning board before they bought the 183 property, and the board said the second lot was buildable. So they bought. Now they want to move on to a larger home, and to finance it they decided to sell the other lot. In comes the conservation commission and says, oh no, that's a wetland, you can't build there. Let's say the lot was worth something like $50,000, that means the state has come in and taken their money. Not fair. It's the kind of thing that starts revolutions. It's not what the Founding Fathers had in mind: take no property without just compensation is what they said—you can't take property without paying for it. Fifth Amendment. Go look it up. I mean that's the law."

I went into the city a few days later and found the Solicitor at his pub. I told him about Farmer Flagg and his wetlands story.

"He's got a point," the Solicitor said. "Up to a point. Can't take it without due process, A, and, B, can't take it without just compensation. Old argument. But he should read his legal history. Good Justice Holmes, God bless his poor benighted soul, beat him to the punch. Property rights, as he called them, are only enjoyed under an *implied limitation.* They have to yield to the police power. If the regulations go too far then it's a taking of property. So says Justice Holmes. Go tell your Farmer Flagg and see what he says."

"I know what he'll say, but what's *too far?*"

"Anything, according to your farmer. But what it comes down to is who picks up the chit? Does the burden fall on a single individual, one man or woman, whereas the benefits are shared by many? Or is there a sort of rough equality between the burdens of the regulation and the benefits? He might not be allowed to fill his wetland, but the burden isn't falling on a single person—law applies to everyone, doesn't it? And the benefits are shared by everyone. Case closed."

"Ah, but Mr. Flagg would argue that the big landholders such as himself are subsidizing the rest because they stand to lose the most if they can't develop their land. They are the only ones who lose, and therefore they alone are subsidizing the greater good. And anyway, wasn't there a case around here, a state case, where the judge ruled that the developer was unjustly deprived of his rights by a wetlands regulation?"

"*Lopes v. City of Peabody.* Not quite a takings case, although the developer won. They're all citing another case, *Lucas v. South Carolina,* where this bloke, this big-time developer, comes in and tries to build houses on a beach.

"A new state regulation had passed that prohibited permanent structures on beaches and Lucas argued that the regulation had destroyed the economic value of his property. He wanted just compensation. Went all the way to the Supreme Court, and the court agreed: state had, in effect, taken his land. Bad day for environmental types.

"But your farmer friend—not to mention Mr. Lucas—is ignoring the fact that he is going to benefit from the restrictions in the long run. So is everyone else in the town. Restrictions enhance the value of everyone's property, don't they? Can't do things that damage your neighbors. Benefits are equally shared by all property owners. Holmes would say that is a general burden: everyone is subject to the same general restrictions. The good Justice would say it's only when regulations go too far that you have to compensate. Supreme Court ruled that South Carolina had gone too far."

"Mr. Flagg would say that his family bought the land in, I think it's 1834, and that at the time he could have done what he wanted there, so . . ."

"Not true," he said. "Always have been restrictions. This is common law, isn't it? Based on precedent. Goes back to Henry II, who standardized local laws. Can't run your ducks and geese in the common stream. Can't throw your slops in the stream between the hours of six in the morning and eight at night. Go back and tell Mr. Flagg that regulations controlling the use of land go back to the twelfth century. English law, don't you know, which your people, in their wisdom, chose to adopt after you so ignominiously threw off the Yoke of Tyranny.

"Did you know there was a brief period after the Revolution when your forefathers were casting about as to how to write their laws when they considered taking on the *French* system?"

He paused here, drank, and rolled his eyes heavenward. "Roman law is the root of civil law in Europe, all based on statutes, written codes; *le code penal,* the Napoleonic code, in the French system. Some states, such as Louisiana and Vermont, adopted constitutions that explicitly rejected some of the English laws," he said. "They were trying to get away from the privatization of public resources. Which is another thing that the property rights people fail to talk much about—greed—private individuals exploiting public resources. Roman law holds that there are certain resources like the air, water, the sea, the land under the sea, which should be commonly shared. Your private property rights types have been trying to limit those public rights and public resources and privatize them. Comes up in the West over grazing rights on federal lands, mineral rights on federal lands, logging on federal lands."

"The Sage Brush Rebellion folks who want to privatize the public lands of the West would not agree," I said.

I had recently come back from a trip to the empty spaces of the Joshua Tree national park in southern California and, in the emptiness of the American wilderness, had once again been impressed with the fact that, in this case at least, landscape, the vast, free, untrammeled wilderness, must have had at least some influence on the American spirit, the laissez-faire attitude of everyone from the robber barons to the present-day right-wing, hands-off-my-property terrorist groups of Montana and Oklahoma.

"Have you ever looked at a map of federally held lands in the United States?" I asked. "There are a few pockets here and there in the East, but in the West, practically all the land outside the cities is held by the feds, Bureau of Land Management, National Forests, and the like. If you lived out there with thousands of undeveloped acres around you and made your living cutting trees or running cattle, you might disagree too."

"I am not now, nor will I ever be, a cowboy," he said.

"Evidently." He was dressed to the nines that day, a dark wool Bond Street suit, red silk handkerchief in the upper pocket, starched white shirt, and a lawyerly necktie.

"Besides, these Sage Brush folks don't know their history," he said.

"Americans don't like history," I said. "This is the New World."

"Tell me something," he asked. "Are not most of your cowboys of the Sage Brush Rebellion of the Christian persuasion?"

"They are."

"And do they not, therefore, attempt to abide by the teachings of Our Lord?" He was taking his courtroom stance.

"They try, no doubt."

"Well, Jesus rejected private property."

"Yes, but his Father told us to be fruitful and multiply and have dominion over the earth and subdue it."

"And replenish it, if I recall. And be stewards of it. 'The earth is the Lord's and the fullness thereof . . .' Doesn't sound like private property to me. Sounds like it all belonged to God, which is what your Indians believed, if I am not mistaken. How can you privatize what's God's?"

"To be a steward you must own it."

"Not at all. The gamekeeper is a steward. The peasant who tilled the common fields of England for a thousand years was a steward. They didn't own the land."

"Your king owned it."

186   "And allowed for common fields. Anyway, the earth abideth forever,

whereas the days of a life are numbered and pass swifter than the weaver's shuttle."

"Yes, that's why we have inheritance laws," I said. But in the end there really is no sense in arguing with the Solicitor. If he can't win, he'll just wear you down.

"Have another drink," I said. "Do you know what the Mole said?"

"Ratty's Mole, you mean?"

"The same. He said we must stick to the places where our lines are laid out."

"What's your point?"

"Well, Ratty and Mole were fixed in their place, the riverbank, the water meadows, the tilled fields and plowed furrows. They fitted themselves into their natural environments without altering things."

"I think you've had too much wine. Tell that to the judge. They're just animals. No standing, no legal authority. You've gone 'round the bend with all these mole and salamander arguments."

⋯⋯ ⋯⋯

The question of private holdings as opposed to the ancient native system of the *ejido* is still very much a part of the land-use disputes in the American West. In San Luis, Colorado, for example, which bills itself as the oldest town in the state, there has been a running legal battle since 1964, when a newcomer to the town obtained, through possibly devious means, from an older local family, the title to a mountain known locally as *la herencia,* the inheritance, or patrimony. The tract, which consists of some 1,700 acres, bears vague resemblance to Nashobah inasmuch as it sits above the community, is an excellent resource, and is dominated by a manor lord. In this case, the lord is a rancher named Zachary Taylor, whose father bought the property in '64 and promptly excluded the locals who had been using the area as common land since the mid-nineteenth century. Under an 1844 land grant the local people of the community were permitted to continue to graze animals, collect firewood, hunt and fish—essentially the same rights that would have been granted, allodially, to English peasants before the time of William or to the Mexican natives before the hacienda system excluded them.

Essentially, the story of San Luis is the story of many old towns in the Southwest. After the Treaty of Guadalupe Hidalgo in 1848, which granted the lands to the United States, robber barons moved in, claimed the land as 187 private property, and proceeded to strip its natural resources, whether timber, minerals, or grazing lands. La Sierra, the mountain now owned by the Taylor family, had been reserved as common land by an enlightened early settler in the region, a Frenchman who wanted to keep the land open to provide for the needs of the local people.

Relations between the Taylor family and the natives declined, and finally the locals sued, claiming their rights to the common. The Taylors actually tried to sell out after the death of the old man, but since the case was by then in court, no one wanted to risk a purchase. So Zachary Taylor, in classic baronial style, called in the loggers and began selling off the timber. It soon became one of the largest timber operations in the state, and the eroded hillsides began to clog the local *acequias,* or water channels, that the people traditionally used to water their crops. Noisy lumber trucks began plowing through the small town at ungodly hours of the night, and the wooded hill, the common resource for the village, began to show ugly scars. All this attracted a grassroots action group that calls itself Ancient Forest Rescue, and the battle raged on. Nothing new here, but it helps put in perspective the comparative benevolence of Sir John Morrison, who, if anything, as private landholder, has restored the landscape of Nashobah.

Morrison is not alone among those freeholders of America who, although fanatical about their rights as property owners and their fear of government controls, are in fact excellent stewards of their land. There are owners of timber tracts in the Pacific Northwest who have had these forests in their families for generations and who tend them and protect them as carefully as an Italian hill-town gardener. A few years ago, during the spotted owl controversy in the Northwest, some two hundred of these land-owning stewards received legal notices from the federal government informing them that it was not legal for them to cut trees on their property. One of them published a simple, heartfelt response to the ruling in the *New York Times.* "Do private property rights still have meaning in America?" he wondered. "Do they understand that the right of ownership of private property is fundamental to our democracy?"

He had a point. From the first hard years on the American continent, the Puritan fathers and the southern tobacco growers believed above all in independence and the right to do with their land whatever they liked. It would have worked had they been good stewards from the beginning instead of dog-eat-dog capitalists bent on private gain at public expense.

A few years before Christopher Stone wrote his essay "Should Trees

Have Standing?" the ecologist Garrett Hardin wrote another seminal essay called "The Tragedy of the Commons." His thesis, which was primarily an exposé of the dangers of overpopulation, used the traditional common as metaphor. Hardin's essay put the case that one farmer, realizing he could make more money by running more cows on the common land, increased his herd, thereby increasing his profit. The next farmer, seeing his success, did the same, and so on, until the common was overgrazed and useless to all.

The property rights advocates have used that essay repeatedly to the point up the errors of maintaining large tracts of public land. But the traditional common always had built-in restrictions: rotation of cropland and pastureland, for example, and limits on the number of sheep or cows that could be grazed in an area. If the common had endured on the small, workable community scale that traditionally characterized these communal open spaces, the system might have lasted another thousand years.

Those of us who live on the outskirts of developed areas, surrounded by private woodlands and fields, are often shocked by little indications of things to come. Walkers such as myself, who normally take the untracked routes through private properties to get to isolated meadows and woodland tracts, plow through some thicket, climb over a wall, crash through a swamp, and climb a slope, only to find there, in the middle of the wild meadow, a yellow backhoe. You wander off into the forest without any idea of where you are going or where you will end up, hoping to become lost, and there in among the trees you see a strip of ruined earth, a track, which you follow until, at its end, you see a pit or a refilled hole looking for all the world like a fresh grave site—which in a way it is.

In the Nashobah region, according to local and state health codes, tests to determine the level of the water table must be carried out in April, when the water is normally at its highest. And so after the first of the month, I go around with trepidation, eyeing every backhoe I see ranging along the country roads in these parts, wondering where will it stop to dig. It's a sort of deadly roulette game. Such tests are the beginning of the end for the open spaces of this region, the prelude to development. On their heels will come the little announcements in the local papers and gossip in the town grocery stores—yet another housing project to go in at so-and-so's meadow or in the woods over by such-and-such a ridge. This is not an isolated event; it is the story of the haphazard settlement patterns of the whole of America, 189 a tendency to move outward into open spaces for new housing rather than develop more intensely the village centers and let the agricultural and wild lands remain untouched.

There are some signs, though, as at the New View development, that this disparate, separated, one-house-one-lot model that has characterized the suburban sprawl in the decades since World War II may finally be weakening. According to Dick Newton, one of the leaders of the fight against (or, more accurately, the fight to control) the Beaver Brook development, people don't want the big manor anymore, with the intensive-care lawn and all the boring chores and the isolation that is so much a part of suburban life.

"This is the sandwich culture," he says. "Two working parents headed out each morning a hundred miles in opposite directions to get to their jobs only to return at night, eat, sleep, and do it all over again the next day. No one really wants that. No one really wants to spend the weekend mowing the lawn and taking care of a big yard. They want to hike or bicycle, sail, or something. You spend twelve hours a day away from your house and two days at home on the weekend. People want to live in smaller houses, closer together, with common land they all can share and let someone else do the upkeep."

He may be right. There is even a name for the phenomenon now—always a sign that something is beginning to change—"the new urbanism." Although the new direction has its critics, who claim the whole thing is a setup by developers, and even though massive trophy houses continue to be built, the movement toward consolidation is being led by planners and architects, as well as developers. One of these new projects is a prize-winning development in Florida, in which the planners recreated an entire nouveau New England village, with narrow streets and picket fences and a meetinghouse. Critics point out the only people who live there are retired southern golfers, but at least the idea of a village is there. A similar idea was put to the town of Mashpee on Cape Cod, where, in what I believe must be the only case of one developer destroying the works of another, builders actually proposed to tear down a strip mall and redevelop it into a small village center with shops, narrow streets, apartments above the shops, and large houses surrounding the center. All of this was contrary to the local zoning codes, which dictated, as most zoning codes do, that development would spread ever outward. Innovation, land protection, and conservation, according to this model, depended on large lots, the larger the safer.

Large-lot zoning does not take into account that there are poor people in this country who can't afford to buy the land, let alone build a new house.

But then, what else is new? In any case, zoning by lots encourages the dreaded suburban sprawl rather than concentration. Unfortunately, the model of large-lot zoning, whose time is about to come to an end in some sections of the country, is very much alive in the West, which is currently experiencing a sort of third wave of outward growth. For all the myth of the West, the wide open spaces and the freedom to roam, a high percentage of Westerners live in Los Angeles–style cities, house to house and lot to lot. Only thirty to forty percent now live in the country. Some of these western cities are experiencing all the horrors of LA–bad air, gridlocked traffic, huge gulfs between the rich and the poor, gated communities, and gang violence. Phoenix actually has more sprawl than the mother of all sprawl cities–some 469 square miles (as of this writing). Los Angeles in fact is working on models to centralize its infamous sprawl. The outward growth is not limited to dry, desert regions of the West, however. Some of the richest farmland in America is suffering the same fate. The American Farmland Trust estimates that as many as 25 million acres of prime agricultural land are threatened by urban development, much of it in key farming states such as Illinois, Texas, and Florida.

The use of zoning began as a tool to control abuses of the land by ruthless developers, and it was quite naturally opposed by those who stood to profit from a laissez-faire approach to building. This opposition led to one of the more significant legal decisions as far as property rights are concerned, the Solicitor says.

In the 1920s communities began using zoning codes to control land-use patterns, mainly to increase lot sizes. There was nothing new about zoning. The Puritan fathers attempted to do the same thing, albeit in the opposite direction, when they passed laws forbidding anyone from settling too far away from the meetinghouse. But in the '20s, one of these zoning regulations, which, as you can imagine, were anathema to real estate developers who wanted to crowd as many houses as they could on whatever land they owned and thereby increase their profits, came up against the court in the village of Euclid, New York. Ambler Realty took the town to court for its zoning laws, and after many small, stiffly fought legal battles, the case went to the Supreme Court, which upheld the right of the town to use zoning codes.

The people at New View spotted the trend toward a new urbanism years ago and designed their minivillage with the houses close together, the meetinghouse and common land in the middle, and more common land beyond the houses. After eight years of planning, after interminable meetings, after interminable explanations to the various boards of the various towns in which they tried to place their common vision, after all this, they finally got approval from the town of Acton to build their dream. It is not insignificant that the planner for the town was at that time a European man. The concept did not seem all that innovative or outrageous to him; he had grown up with it. What's more, there is a certain irony in the fact that the site where the New View people planned to build was no more than two miles away from the last communal, co-housing experiment in the area, the Christian Indian village at Nashobah, disbanded by force some three hundred years earlier.

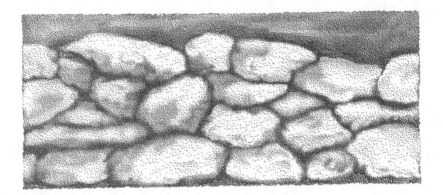

Chapter Fourteen

The Council judging it of absolute Necessity for the Security of the English
... Do Order that all those Indians that are desirous to Approve themselves
Faithful to the English, be Confined to their several Plantations under-written,
until the Council shall take further Order....

—Order of the Authority of Boston, August 30, 1675

The Tawny Vermin

I woke up one morning in early February and heard the woodpeckers drilling the trees in the woods beyond my garden. Sun streaming in the bedroom window: the song of cardinals, the whistles of chickadees and titmice. Spring, a process that in this part of the world can go on interminably, had made its first wedge in the solid block of winter. Birdsong notwithstanding, there had been a light snow the night before, and I thought this would be a good day to look for tracks at the village site.

I was down at the tract by early morning with a thermos of coffee, which I carried up to the high ledges above the serpent mound. I found a sunny spot there and sat down to drink coffee and think things through before I began my tracking adventure. There was a good smell of warm earth in spite of the fresh snow, a hint of tree bark, of swamp, and fresh, clean air. Sitting there, overlooking this wooded landscape, quite naturally I started to think of the Indian hunters who would have been so hard pressed at this period of the year. Oddly enough, now, in this crowded, suburbanized part of the world, a few of the species that were already extirpated from the region in Sarah's time have come back. The beaver, which was trapped out of the region early in the seventeenth century, has returned with such force as to become a pest to those living in low-lying areas around streams. The coyote, which seems to have moved into the region in place of the wolf of Sarah's time, is common in the woods. Moose often pass through, although they would have been fairly common when Sarah lived here, and the deer, which were totally extirpated by the end of the eighteenth century, are also common. Bears are coming back too, although so far they have yet to make it as far east as Nashobah. This of course is a sadly diminished list. The world was far more diverse in her time. Immense seasonal flights of passenger pigeons would darken the skies for days, there may still have been a few woodland caribou in these parts, the bear and the wolf were common, mountain lions and bobcats may have denned in the very ledges where I was sitting, and the streams were filled each spring with such thick runs of tomcod and herring that their backs glistened in the April sun.

I was thinking these long thoughts about the past glories of the wild populations when, in one of those strange coincidences that often occur in moments of daydreams, I looked down at my feet and not three yards away saw an arrow. It was, unfortunately, one of those hideous modern steel or alloy arrows with a cruel barb at the end and some kind of plastic that served as the feathers, and not a mystic hickory arrow flown in from Sarah's time. Still, it was a stark reminder of the reality of the hunt, the downright bloody cruelty of the strike and the fall and the kicking legs and gasping jaws. It's no wonder the natives of this region went through such highly ritualized propitiations at such times.

It was a good day for tracking, as I expected. Down on the lowlands I saw the large chickenlike footprints of a group of turkeys wandering along, and then, interestingly enough, noticed that they were joined by a set of coyote tracks, which came up to them at an angle and then turned and followed the turkeys. At one point the turkey tracks ended abruptly and the coyote tracks wandered off. Clearly, the turkeys had taken to the trees. I searched the branches but saw nothing but the skeletal winter oaks, their remnant leaves still hanging like bat clusters.

But it was a fine day withal—squirrel prints and rabbit tracks and the single line of a fox who wandered through. Here a grouse, here a little sewing-stitch mouse trail coming down from a tree, winding along the ground through the overhanging limbs and then disappearing into another tree. A raccoon track at one point, another coyote, or the same one, still hunting something, and then a set of prints I couldn't identify, the larger, bounding pattern of some tree-climbing mammal, perhaps the elusive fisher, which I have heard is returning to this area. One woman I met in the woods here told me once she saw a monkey, or a thing like a monkey, high in a pine tree near this spot. The more I asked her about it, the more I became convinced that she had seen a fisher.

Little oak galls were hanging from the bare limbs. I found the gray shielded beetles I've always mistakenly called soldier beetles parked in the crevices of south-facing trunks of trees. I found a pepper scatter of snow fleas at the base of another tree, and down by the little brook that snakes below the head of the serpent mound, I found a stone fly resting on a sunny rock. Then below me in the swamp, poking their green noses up through the wet mud, I saw the brave shoots of the skunk cabbage, the earliest emergent plant in these parts.

Nature, contrary to the myth, does not sleep all winter. It gets up on sunny days and stretches itself. But that is not to say this is an easy season for those who do not hibernate or migrate. Like the other local mammals, Sarah and company would have been facing hard times at this point in the natural year. Corn supplies would have been stressed. Runs of fish were a month or so away, green shoots of early edible plants would not be up until April, and the ground would be too frozen to dig for roots. Even the hunting would have been sparse. Certain bands of Indians in the region had a technique of marking bear hibernation spots with a stick. In midwinter hunters would search for holes with yellowed patches of snow, a sign that a bear might be sleeping beneath. They would then mark the spot, and in late winter and early spring return to dig the bear out. Whether this happened at Nashobah is not recorded, but there is no question that Indians all over the north were stressed by winter. It was one of the things that shaped their cultural traditions. One friend of mine, Lawrence Millman, who has lived among the last hunter-gatherer tribes in North America in Canada and Labrador and has translated their folktales, says that the hammer of winter has forged the traditions of the tribes even as far south as the mid-Atlantic.

It is unlikely, however, that Sarah complained about the gnawing hunger of late winter and the shortage of corn and greens. This was, in her mind, the way the world worked, and she lived with the fact that all the other living things around her were going through the same process. In effect, even after her conversion, she probably did not think of herself as separate from the chipmunks who emerged on sunny days in February and scurried through her village, or the woodpeckers that began their frenetic tapping in February. The oneness of the Native American cultures with the earth is almost cliché, but as far as our story of land ownership is concerned, it was critical: survival depended on the shared resources of the common ground.

On August 30, 1675, in response to the early success of King Philip's uprising, the Council of the General Court, a group of some twenty-four deputies elected by adult male members of the church, issued a proclamation that suspended all the rights—such as they were—of the Indians of the Massachusetts Bay Colony and ordered that all be interned in camps. After that day, any Indians found roaming beyond the bounds of the designated areas were to be considered dangerous and shot on sight. Most of the Indians in the established praying towns of the colony, including the Nashobah people, were forcibly removed from their legally granted lands and transported to Deer Island in Boston Harbor.

The proclamation effectively ended the missionary work of John Eliot, although for the rest of his life he continued in his efforts and even stepped up his defense of Indians during the dark years of the war, when all his converts were rounded up and taken to the bleak rocks of the island.

On a late autumn day in the year 1675, the women of Nashobah were out in their gardens or perhaps scouring the lands around Nashobah for acorns and beechnuts and chestnuts. One of them, or someone in the village perhaps, heard approaching horses and the regular tramp of marching men and saw forms moving between the trunks of the trees near the Great Road, where the main Indian trail ran. In time Captain Mosely appeared, his slouched Monmouth cap turned back and clipped with a silver pin. His scurvy troop of pirates came on behind, some perhaps on horseback, most of them walking and armed with muskets, swords, and pikes, and a few wearing the breastplates of Oliver Cromwell's time and rounded pot-shaped helmets with flaring, upturned brims. They turned at the southeastern end of Nagog and cut through the area known as Crow Hill, skirted the southern shores of the pond, and headed for the village, perhaps passing Indians on the way.

The English drew up at the village, a loose circle of bark-covered wigwams and perhaps a pole-framed English-style house or two. The women were tending fires, the men lounging, chipping points, eating from the common pot, dogs charging about, children running out to see the English, and most especially, no doubt, to see the horses, which were not common in seventeenth-century New England. Great excitement, but no real alarm. And in he rode, a stranger, not Gookin, whom they knew, not John Eliot, whom they must have trusted—someone unknown to them. But nonetheless, when Mosely came in they must have come forth and said in so many words, "Welcome, Englishmen." In short they still had trust. They were Christians.

Then Mosely delivered his ultimatum—in English no doubt, a language that was only partially comprehended by the Indians at Nashobah, although someone there on that day must have spoken English, Sarah herself for all I know, but she would have held back with the other women when she saw the English, having by that time had a taste of their ferocity. She was still in mourning for her son and her wounds were not yet healed.

Mosely shouted out commands in his broad Kentish accent, words that we, in our time, would have trouble understanding. His a's were flat: he pronounced the word "war," for example, as we would pronounce "ware" as in "This be toime of ware."

Possibly there were no words at all. When Mosely rounded up the people at the nearby Christian Indian village near Marlborough, he used violence to make his point. He roped the people by the neck to keep them from running away and marched them—women, children, old people, and fighting-age men—some forty-five miles through the winding woods paths to Boston. So at Nashobah they may have stormed in. They may have played out a prelude of events that would take place over the next two hundred and fifty years on the American continent and have carried out the traditional predawn raid. That is to say, they may have charged into the village firing, scattering the women and children, burning the Indian shelters, and killing the dogs. Mosely was more intent on capture, though. Perhaps his men ran their halberds through the bark or skin walls of the wigwams and shouted for the heathen to come out, while inside, the Indians crowded together, the children crying, the old perhaps reverting to their ancient religion and singing the Algonquian equivalent of a death song, believing that the end had come. Perhaps the English sounded a trumpet. However they did it, they collected the Indians there in the center of the village and then ordered them to take what they could carry—one presumes—there is no record of this either way except that it happened at some of the other Christian Indian villages, where the people were given an hour or two to collect goods. The Indians, as I have said, having no inclination other than loyalty to the Crown and Christ and fully believing in the courtesies and benevolence of the Christian god Jesus (god with a lowercase as I think they must have understood it), would have obeyed. They went into their wigwams and collected their things, the men carried a few tools perhaps, bows and arrows, if they were permitted—which is unlikely—and the   199 women, their baskets and a few articles of clothing. But no food, none of the collected harvest of the summer, upon which these people so depended. And then, on command, perhaps with the encouragement of a sword point, or a halberd head, or a certain amount of unmistakable body language, they would have begun their eleven-mile walk along the narrow Indian path through the deep forest to the English village at the Indians' former planting fields at Musquetaquid, the place the English now called Concord, a town named for peace and accord between peoples.

For some of them, having been born in this place, this was a return. For some, it was new territory. For most, it was a death march.

If all this sounds vaguely familiar–this uprooting of people from lands that they legally hold and transporting them to another venue and making them settle there–that is because it has happened many times since. Nashobah was merely the first. The last (at least one hopes it was the last) also occurred in time of war.

On February 19, 1942, after a certain amount of legal debate, the liberal-minded president of the United States, Franklin D. Roosevelt, signed a document known as Executive Order 9066. The order gave the Secretary of War the authority to designate certain inland "military areas" in the West, where Japanese aliens and American citizens of Japanese blood, that is to say most people of Japanese ancestry, should be relocated to internment camps for the duration of the war. As a result of that statute, on a given morning– any morning between 1942 and 1943 would do–federal marshals would appear in Japanese neighborhoods throughout the far West and inform the residents that they were to prepare to move.

Unlike the exotic, non-English-speaking "salvages" of Nashobah, these people wore American clothes–pinstripe suits and white shirts and ties and shirtwaist dresses and high heels. Most of them spoke English, some of them with no accent, some with no knowledge or memory of their native tongue, and many never having set foot in Japan. They were most of them, in effect, American citizens, *Nisei,* second-generation Japanese, along with *Issei,* the first generation. Regardless of citizenship, of professed loyalties, of pleading or looks of confusion, the evacuation, as it was called, was carried out. When the transfer was completed, some 110,000 people were removed from their private dwellings–many of them owned outright under American law, complete with deeds–and transferred to internment camps.

It all began slowly, as it had with the Christian Indians, with a few raids against suspicious individuals, such as Togo Tana, the outspoken but decidedly pro-American editor of the Japanese-language newspaper *Rafu Shimpo,* who was picked up by FBI agents and held without explanation in jail for eleven days, beaten, fingerprinted, never charged, and then inexplicably released. All that year the process continued—mild-mannered gardeners, arrested by the FBI; innocent men who ran the local grocery stores charged with espionage; cleaners who always smiled and pressed shirts for the local lawyers deemed enemies of the United States of America. And then later, with the signing of the Executive Order, the Gestapo-like roundup, the barbed-wire camps with towers and machine guns, and throughout all of it, the abuse, the land grabs, the houses left empty, the shops boarded up, and all this strictly legal by authority of the United States Government. Much of this dark side of American history, so long absent from approved textbooks, is now coming to be recognized and, in some cases, even acted upon. Survivors of the Nisei internments have successfully sued the federal government for damages.

⬤〜⬤

By the winter of 1675-76 there were some five hundred souls enduring the insufferable conditions of a New England coastal winter. According to Gookin, the wigwams of the Indians constructed at Deer Island were but poor and mean, and the people were not well clothed. They lived chiefly upon shellfish, which they dug out of the sands at low tide, and what little dried corn they had with them. Eliot and Gookin periodically organized expeditions to deliver corn to the people, but the wrath of the English against all Indians was so great that they were prevented from doing so on a regular basis. At one point, Eliot's boat, carrying supplies to the Indians, was rammed by another vessel and capsized, throwing Eliot into the sea. But the good Mr. Eliot, forgiving to the end, said that it was an accident.

Gookin visited the people on the island and reported that although they were living in extreme conditions, they carried themselves patiently and humbly and piously, without murmuring against the English for their sufferings. (Bear in mind that this is an Englishman delivering the report.) Furthermore, this same document says that, by their great suffering, God was manifesting his obscure benevolence. The Christian Indians on Deer Island, Gookin argued, were safe from the fury of the colonists, who would just as soon have strung them up on the local gibbets. Furthermore, God caused by these sufferings a humbling and a betterment of these people. It did not occur to him that Job in all his sufferings had not experienced the hardships that these Indians had undergone since the Europeans first landed on their shores—loss of their planting grounds, plagues, and outright massacres.

The Nashobah people were the last group to arrive at Deer Island. After the raid on the village, Sarah and many of the other women and children of Nashobah were first taken to Concord and placed in a stockaded yard under the care of a man named John Hoar, who had agreed to undertake the responsibility for what he and a few others in the town—but not many—felt to be a harmless, loyal people. Using his own money, Hoar built the stockade and workshop near his house and ensconced the Indians there and set them to basket weaving. According to early histories of Concord, he even under-took a journey to Nashobah with an ox team to retrieve the supply of corn the people had left behind. This was no mean feat. By this time, the winter of 1676, King Philip's followers were closing in on the Concord region. They had burned all but four of the houses in the town of Groton, just west of Nashobah. In late March they burned Marlborough. Lancaster had been decimated, and Mary Rowlandson, the minister's wife, had been taken pris-oner. They had raided Sudbury and killed the wife and children of Thomas Eames, and had made a lightning raid on the farm of Isaac Shepard, just north of Nashobah and within the bounds of the plantation.

Shepard had posted his fourteen-year-old daughter, Mary, on Quagana Hill to watch for Indians while he and his sons threshed the grain in the barn. Either the Indians surprised her, or otherwise silenced her, and got into the barn and killed her father and one brother. Mary was taken prisoner and carried back to a camp somewhere west of Wachusett where Queen Wetamoo and her warriors were gathered. But one night, the story goes, Mary stole a horse, swam it across the Nashua River, and managed to escape. A week later she turned up in Concord, where her mother's family lived.

Tom and Sarah Doublet probably knew this family, since the Shepards lived on the northern boundary of the plantation and no doubt had traded with the Christian Indians from time to time. The family must have negoti-ated the purchase of the property from the Nashobah people.

This John Hoar, who was a cantankerous, independent man, was taking his life in his hands to retrieve the supply of corn. He and Gookin and Eliot, and a few others in the town of Concord, were the Indians' only defenders. The whole community found the presence of Indians threatening, even though the prisoners were mostly old men, women, and children, and even though they were professed Christians. For all this, the Nashobah people were probably better off than any of their brethren, since the August Proclamation had decreed that any Indians found outside their village, Christian or otherwise (of course), could be shot, and on Deer Island, under the bleak winter sea winds, supplies were running short.

According to Townsend Scudder's *Concord, American Town,* on the Sunday after Mary Shepard's return, just as the people were filing into the meetinghouse, a horse company clattered into town with dark Mosely at its head. The troop of brigands listened to the sermon and then Mosely himself spoke to the congregation. Most knew why he was there. Scudder says he had been summoned.

Were the Nashobah people not still here in their midst? Mosely wanted to know. They were. Why? Mosely wanted to know. Why do you harbor these "tawny vermin"?—as the current jargon termed Indians. Will you not have them removed? he wanted to know. Although most remained silent, a few shouted in agreement. And so Mosely went around to Hoar's stockade on the east end of town and hammered down the gate. John Hoar demanded to see Mosely's order from the General Court, which Mosely could not produce, having only a commission to fight Indians and nothing more. Hoar refused to accept his authority and Mosely went off. Scudder says the entire community had trooped out of the meetinghouse to the stockade and stood there, silently watching the interchange.

Mosely came back the next morning, with more determination and, although no history says this, not even the imaginative Scudder, probably a healthy share of ale under his belt. He and his men, like free-lance foot soldiers the world over, had perhaps spent the evening steeling themselves for a good fight. This time Mosely pushed John Hoar aside, and his men fell to stripping the Nashobah people of their minimal possessions. Then he lined them up again, between two files of horses, and marched them east to Deer Island.

The choice of Deer Island as the internment camp seems to have laid a curse upon this piece of earth. The island may have absorbed all the evil that was inherent in the act of transporting loyal citizens, the actual natives of the land, to this wintry place and then letting them fend for themselves on their limited supply of corn and what shellfish they could garner from the rocky shores. Since 1676, Deer Island has been a sort of repository for human outcasts—Indians were buried at the site, and later the island became a cemetery for Irish immigrants and then later, a prison, and still later, a sewage treatment plant. In the past decades, it has become the site of a major revival of

Boston Harbor waters. The Massachusetts Water Resources Authority, after much debate and bad publicity, including a few slams during the presidential elections of two administrations, announced plans to construct a state-of-the-art sewage treatment plant on the island. By the winter of 1997, the project was nearly completed and a secondary treatment plant began cleaning the waters enough to bring Boston wastewater discharge into compliance with federal standards for the first time in history.

But that is not to say that all was finally well on Deer Island. For one thing, wastewater, not entirely cleansed, is discharged into Cape Cod Bay at the end of a nine-mile tunnel. This has stirred a hornet's nest of alarm among Cape Cod residents, who fear that the tidal currents in the area will swirl the effluents around into the bay and in effect export Boston's dirty waters to the relatively clean waters of the Cape. There was also the three-hundred-year-old problem of the burial of the good Christians of Nashobah and the other Indian villages.

Generally speaking, the human population of the twentieth century does not know, much less care, about the events that transpired on Proclamation Day in 1675. Only a small percentage of people have even heard of Christian Indians, and, history courses being what they are in our time, most students have not even heard of King Philip's War, especially those outside New England. But the descendants of the Narragansetts and the Wampanoags and the Pawtuckets and the Nipmucks and the Pokanokets have not forgotten.

In 1994 a group of Indian activists from the loosely organized group of tribes known as the Muhucanuh Confederation banded together to publicize the crimes that were committed on the native people during the war. Recent atonements to the Nisei, who had won a few lawsuits, may have been on their minds. These people see Deer Island and, to some extent, even the Christian Indian villages as concentration camps. Their theory of history—not an entirely paranoid view—is that the racist English Puritans, accompanied by the French and the Spanish in other sections of the American continent, set out deliberately to exterminate the native people through warfare, introduced disease, imprisonment, and the use of concentration camps, a.k.a. reservations.

After some three years of struggle, the Muhucanuh Confederation managed to get the city of Boston to recognize Proclamation Day and officially declare an anniversary. And so, on August 28, 1996, representatives from the office of Boston's mayor and adjutants from the German and English and

Canadian embassies, along with a company of native people from the

Wampanoag, Narragansett, Penobscot, and Stockbridge Indian tribes, gathered on the island nearest to Deer Island to hear a proclamation read and to deliver speeches.

Some of the speeches from the mayor's office and other officials were remembrances of the dark day in 1675 when the General Court issued its orders. Many of the facts in these official orations, white and Indian alike, were erroneous. Thousands did not die on Deer Island, as was claimed. There were hardly a thousand Christian Indians in the whole of the Massachusetts Bay Colony, and only five hundred were imprisoned there. But the whites expressed their dutiful mea culpae, and there were some fine harangues from the assembled Indians, many prayers to Mother Earth, and many sachems and medicine men passed among the assembled with burning leaves of kinnikinnick, the sacred herbs the Indians use to purify a place. In this case it was an excellent choice. For some reason—a few wags thought it may have been revenge—the mosquitoes at that ceremony rose in such abundance and with such ferocity that not a single speaker failed to mention them, and they had the assembled slapping and scratching during the entire ceremony. The smoke of the sacred herbs at least held a few of them at bay.

Some of the Indian participants had the grace to thank the current mayor of Boston for signing the proclamation. A chief called Helping Beaver spoke of the seabirds and the mosquitoes and how we are all one part of the great circle of being. Another man wished to know what had happened to his ancestors, the fourteen million victims of what he called the American holocaust. Many began and ended their speeches with prayers in their native tongues.

At the end of the ceremony, I stayed around talking to people about the event. Some of the Indians had come from as far away as Ontario, and I was curious as to why a representative from the German Embassy was there. After some chatter about mosquitoes and American history, I managed to ask the man why he was sent.

"Germany is sensitive to this sort of thing," he said quietly.

A large, sleepy-eyed man who had seen me taking notes and talking to the various Indians there came over and asked if I was going to write something about the event.

"I don't know," I said. I didn't actually know that I was going to write anything.

"Write about me," he said, winking. "I'm the last of the Mohicans. They call me Nanachuse. Means Big Bear. I was the chief of our people."

"That right?" I said. "The chief?"

He was dressed in a flowered blue shirt with a small collar like those the Navajo wear, and he had a few silver bracelets and a leather thong necklace with a tiny black Zuni bear amulet hanging from it. In some ways Nanachuse looked like he had never been off the streets of Boston. His hair, which he had pulled back and braided into a little pigtail, was long and crinkling, his skin was the color of milk chocolate, and he looked as if he had a touch of Cape Verdean blood. He had an easygoing manner, and a little twinkle in his eye that suggested that he may have had some perspective on this matter of modern-day Indians in modern-day America. So I told him about the Nashobah lands.

"Out by that pond there?" he asked.

"Yes, you know the place then?"

"Heard of it," he said obscurely.

In the end I invited him out for a walk.

⊷～⊶

The day we went, I met him at a nearby restaurant and drove him in my car through the orchards of Morrison land and down the hill to the little wooded pull-off where I used to leave my car. I had left some things in the trunk and had to shuffle around the litter of canoe paddles, ropes, tires, books, and gear that seem to somehow accumulate there to find the compass I was after. When I stood up, Big Bear was nowhere to be seen. I thought he must have gone off into the forest, as bears will do, to relieve himself, and so I waited by the car. He didn't return. I waited some more and then went off to look for him. Oddly enough I found him up in the woods examining a tree about a hundred yards southeast of the serpent mound where in the past I had seen little "donations," as the antiquarians who study this hallowed ground call them. Why he had even walked in this direction was something of a mystery, since I told him we would go first to the pond.

Big Bear was poking around in the soil at the base of the tree. He had found a spider in the leaf litter.

"That there'll give you a nasty bite," he said.

As far as I could tell it was an innocent wolf spider, although I suppose it could have delivered a bite if given a chance. I let the matter drop.

This particular tree had a broken-off spiky limb stump and here in the past I had found sticks and pine cones arranged in such a manner as to point directly up the hill to where the mound began. My journalistic paranoia was lurking in my mind and I thought to ask him, in so many words, if he had ever been here before, and if by any chance he knew who was leaving these "donations" and if perhaps it was even he himself, Big Bear, who was a part of this phenomenon. He denied it all. He said he had heard about this tract but had never been here before.

We walked up along the serpent mound to the hillock, or snake's head. He was strangely quiet on the way and walked slowly, looking around, and carefully catching hold of the snags that I held for him as I passed, so as not to snap him in the eyes. He took it all in a matter-of-fact way. If he hadn't been in this place before, at least he seemed to have been in the woods. He was not dressed for the woods, though. He had on his flowered blue shirt again, and pointy-toed cowboy boots with a heel, and he was carrying a soft drink can with him. I was hoping he wouldn't throw it on the ground and force me to give an anti-littering speech.

At one point he cut off from following me and examined something on the ground. I joined him.

"Striped wintergreen," he said. "*Nonichwow.* Good stuff."

"How do you know that?" I asked.

"I been around. Was in the army," he said, as if that had anything to do with it. "Then I was down to the Four Corners area. Sold truck tires there for a while. You're on the phone all day."

None of this explained why he knew the striped wintergreen of the New England woods, which is not a common plant. But this too I let go. I was more interested in his reaction to the Nashobah tract. But I wasn't getting much.

"Nice place," he said when I showed him the great crenellated granite walls north of the mound.

"All right," he said when I showed him the little grassy mounds at the top of the hill, which some believe are the remnants of the corn hills of Sarah's people.

"Yep," he said, when I took him down through the dark forest of hemlocks and white pine to the shores of Fort Pond on the west.

"Nice place here," he said as we crossed the strange flattened granite outcroppings that were reportedly part of the nineteenth-century quarry operation in this area. "Who owns all this, anybody?"

I explained the division of lands here, the public trust land and the privately held Morrison tract, and the small inroads of yards upon which we had been trespassing.

"Pretty good land here," he said.

We went over the hill to see if Rick Roth was around. I felt maybe the   207 presence of a third party would get Nanachuse talking, but neither Rick nor his family was home. So I showed him the pigs.

"What do you call that thing? Potbellied pig? Vietnamese?" he asked.

"Yes. Vietnamese. Were you in 'Nam?" He was about the right age.

"Nah," he said.

We crossed over the wall and descended the hill through Morrison's orchard. I was hoping the old man was not out in his golf cart. On the other hand, it occurred to me, an encounter with Morrison might liven things up a bit. It would be interesting to see how this seemingly lazy, monosyllabic individual would react to the indignant outburst of the Raging Bull. Bear and Bull, so to speak. On the way down I saw Dennis fixing a ladder to a tree and went over to talk. I introduced the two men, formally—on purpose. Neither was a major talker.

"This is Nanachuse," I said.

"Nanachuse," Dennis said with a finality that implied a certain incredulity.

"Means Big Bear in his language."

"Big Bear," Dennis said with the same intonation.

"That's right," Nanachuse said. "Real name is Jim Baggins."

"Well, Mr. Jim Baggins, how do you do?"

"Doing pretty good. Nice place here."

"It's all right."

"You working?"

"Right. Seven years."

"Nice place."

"Right, mon."

"What's the man like?"

"Oh," said Dennis. "The old mon's all right, stay out his way."

"Hiring?"

"You come back in September. Might be," Dennis said.

He looked away, down through the ripening Baldwins to the blue water of the pond below us. An old farm truck, spitting and coughing, chugged up the hill and pulled into the driveway. Dennis hitched up his dark wool pants and pushed the wool cap back on his head as he watched it. I knew him well enough by now to know he was looking for a way out, so I said we had to be off and we started to say good-bye.

"This Nanachuse," Dennis asked as we were leaving, "what kind of name would that be? What language is that, by the way?"

"That's Indian."

"Indian? You Indian?"

"That's right. Mohican. Used to live around here."

Dennis merely nodded and said nothing.

We walked down through the trees, crossed a wall and hiked through the woods to the hillocks of the serpent mound area, and then crossed the road and went out onto the peninsula that juts out into the pond. We sat down on a flat rock by the water.

"You know that black guy back there?" he asked.

"Yes, he's the foreman."

"What's he, from the Islands?"

"Jamaica."

"Jamaica. I been there. Nice place."

"What, when you were in the army?"

"Nah. Working."

He got up and walked back into the woods and then came back and handed me a twig of black birch.

"Eat that," he said, putting a twig in his mouth.

The twig had a strong taste of wintergreen.

Nanachuse was not in any way unfriendly during his monosyllabic afternoon. He seemed to be enjoying himself in a sleepy-eyed, summery way. It was hot that day, the sort of dry, late-summer heat that settles in behind your eyes. We could hear the high whine of a cicada in the trees on the hill behind us, and a banjolike twank of a green frog sounded out from time to time from the shallows behind us.

"Have you spent a lot of time in the woods?" I wanted to know.

"Yeah. Pretty much."

I waited. Maybe he just needed time to collect his thoughts.

"What work were you doing in Jamaica?"

As soon as I asked, it struck me that it was an incredibly stupid question. He could have been carrying drugs and then I would be forcing him to lie, and I had only mentioned it to make conversation.

"I worked on a boat. Big yacht from City Island. Just cooking. You know. Feeding rich people."

"Ever spit in their food?"

I don't know why I asked that, it just came out. I was getting to the point where I didn't give a damn.

"What?"

"I worked with a guy who cooked on a yacht for a rich man. If people on board did something he didn't like, he'd spit in their food."

"They say slaves used to do that all the time. Worse stuff, too," he said.

"This same guy served his people a barley soup with insects in it one evening," I said. "Later one of them came to the galley and stuck her head in the hatch. He figured he'd been caught. 'Delicious soup tonight,' she said."

"That right? Well I knew this guy. He was a logger, lived out in the boonies, in a cabin, used to keep a pot of stew on the stove all day. One year some guy kept coming into his cabin and eating half his stew. So like one day he finds a dead cat out on the road. Puts it in the stew, fur and all, covers it over with sauce and potatoes. That night he comes home, right? Looks in the pot and there's this cat leg stickin' up out of the stew and a line 'a barf leadin' out the cabin door. Never got bothered any more.

"But we had some good times when I worked the camps. Good money too. I just got sick of the work. I got to thinking, what am I doing cutting down all these trees? I don't care how good the money is, so after that I quit. I came back east."

Finally stories of his life emerged. He was born in Wisconsin, I believe on one of the reservations, although he didn't say where. He never thought of himself as an Indian, except that his grandmother used to tell him stories about these people, her people, and their long history. It wasn't until he was older that he realized that the people she was talking about were Indians. He did some family research and learned the story of the Stockbridge Indians, who for a while got caught in the middle of the wars between the English and the Indians and the French. The Stockbridge were actually the enemies of everyone, or at least were a sort of nonallied nation, but for their own protection they had thrown themselves in with the English, then the Mohawk, then the Iroquois. In the end the whole group of them got transported to Wisconsin by the English, but the older people never forgot their ties to Stockbridge, Massachusetts, even after two hundred and fifty years. Nanachuse heard about the celebration at Long Island to honor the Algonquian-speaking peoples who had died there, and since he figured he was one of them, in a way, came east with a group to go to the ceremony where I had met him. He was indeed one of those men who had done nearly anything–within the law; he was a law-abiding citizen apparently– to keep himself alive: worked logging camps, joined the army, learned computer sales, sold tires, cooked for the camps, cooked for yachts he'd pick up in Florida, worked as an extra in some cowboy-and-Indian movies, worked as a wiper on a container ship, a janitor in an American resort in Mexico, and on and on.

"I spent a year reading in the New York libraries. Wanted to learn about Indians, since I'm supposed to be one. But I'm no chief. And anyway, this Indian thing. We're just people. You could be an Indian. The Indian is inclusive, not exclusive, see what I mean. You want to be an Indian, join the dance."

"What about this site, here? Do you think this is Indian land?"

"Could be. I heard talking about this place back after the ceremonies. There was a village here is all I know. Good spot though. Hey. It's Indian land, right? Used to be anyway. Every place was."

Chapter Fifteen

As much land as a man tills, plants, improves, cultivates, and can use the product of, so much is his property. He by his labour does, as it were, enclose it from the common.

–John Locke, Second Treatise on Government, *1690*

Their Heirs and
Assigns Forever

I set out one March day to walk over to the site from the center of town
and crossed, en route, the land belonging to Mr. Couper. The lion
brown fields behind his house were just warming after the winter; you
could see the little mouse trails running through the long, snow-flattened
grass, and the hummocks and hollows were still oozing with winter melt-
water. In a section of old apple trees, just to the north of the field I was in,
I saw a bearlike form under the trees moving slowly on all fours over the
ground. Slowly this bear emerged into human form on its hands and knees,
and I saw that it was no less than Mr. Couper himself, creeping through
the trees.

God, I thought, he's gone around the bend, finally, and transformed himself into one of his beloved animal allies. As I got closer, I could see that he
was patiently parting the grasses with his hands as he crawled.

"God-awful mice," he said. "They'll girdle every damn tree in the universe
if they get a chance. The Lord should have created more snakes and fewer
mice if you ask me."

Like many of his age, Mr. Couper is a great raconteur of past events.
Unlike some of the older farmers in this area, however, he hardly requires
drawing out and is effusive in his accountings, giving details in such profusion and with so many sidebars, and meanderings, and names of those long
dead, that it is sometimes difficult to follow his stories or even to determine
what century he's talking about.

While I stood talking with him under the trees, I was eyeing the back of his old hulk of a barn. It was an angle I was not used to, and I was admiring the view. Mr. Couper saw me looking and told me that some time ago a whole wagon had fallen through the floor of that barn, carrying with it his truly and a good dog to boot. The two of them were half buried in cow manure and had to be hoisted out by his passing father, who, Mr. Couper explained, "was of a mind to leave me there to cook for a while." It was then that I realized that this event had taken place not this past winter but back in 1924, when Mr. Couper was twelve.

That would have been about the same time that Virginia Mills drowned in Nagog Pond, the same years in which the dark-haired queen of the Isle of the Dead would spirit across the pastures and orchards on her white horse. It was also during this period that Mr. Couper's cousin, Chicken John, would have been working on his father's dairy west of Nashobah and that the old patriarch, Francis Flagg, still knew where to find the springs and the old foundations of the Indian fort that had been built at the Christian Indian village site to save the Nashobah people from the raiding parties of Mohawks.

In the twenties, the site of the fort was known as Speen's Point, named for John Speen, a contemporary of Sarah's, and one of the last survivors of Indians associated with Nashobah. Mr. Couper and his cousins knew the area around the point well. He would draw water from the spring, he told me, when they were haying the back fields of his uncle's properties. That got him onto the subject of haying, which got him onto the subject of pasture-land, which got him onto the subject of farms and the shortage thereof in this fast-developing town, and that got him onto what was apparently a favorite memory of his—the great spring drives of heifers and dry cows up the Great Road to the high pastures of New Hampshire.

Farmers in the town would assemble their cattle in a selected pasture and then, herded by the local swains and, according to Mr. Couper, a few hardy dairymaids as well, the whole band would move up the rural road to the town of Ashby, where the farmers had rented pasturelands. The journey, which today can be done in forty-five minutes or so, took two or three days. The swains and the dairymaids would camp along the way in road-side fields, and these were, said my informant, as fine an opportunity as one had to get away on vacation for a few days. Other old-timers in the Nashobah region used to tell me about these drives with equal relish. And indeed, since most of the drivers were teenagers, they must have been spirited events. In some ways the drives were the last of the communal efforts that used to take place in farming communities—shared lands and shared work.

Some drives came in the other direction, when sheep and even turkeys were brought down to the Boston markets on the hoof rather than in trucks. One of the things about the drives is that they would hold up traffic. The drovers would do their best to get the herds over to one side of the road or the other to allow cars to pass, but, inasmuch as cows are mindful creatures, sometimes long lines of black Fords would fetch up behind them and crawl along at a cow's pace—or in autumn, a turkey's pace.

These roadblocks are a phenomenon that still occur in rural Portugal and northern Scotland. But there is a twenty-first-century version even around Nashobah. Every June, immense, overloaded hay trucks, stacked high with the first cutting, pull out onto the Great Road and rumble along at twenty miles an hour, spilling hay as they move. This parade, which often takes place later in the day, after the hay is loaded, often holds up the speedy little cars of computer company workers. You will see them there, pulling out into the left lane to get a view around the truck and then ducking back in at the last minute like fighter planes.

There was a rumor that Mr. Couper, who is not known for his speed on the roads and drives a very old white Buick with broken back windows and a dragging muffler, chooses to drive to town at rush hour at his appointed rate of speed, that is to say, twenty miles per hour in a zone marked thirty-five. The young, fast-lane computer workers who otherwise take this stretch at fifty, sometimes more, line up bumper to bumper behind him.

One member of the Nashobah jester chorus, having witnessed this phenomenon on several occasions, has even named the odd parade: "Couper's Choochoo."

"All gone now," Mr. Couper said when I asked him about some of the people who would go with him on the cattle drives. "Except for Chicken John and a few others. Most all gone."

I mentioned that I had spoken with Chicken John not long ago.

"Well, he'll be gone soon enough," Mr. Couper said. "Him, Morrison, Junior. We'll all be gone soon enough. And what's to become of this world then?" he asked himself.

Mr. Couper meant Nashobah, I thought, not the world at large, which would no doubt carry on with or without Mr. Couper or, for that matter, any of us. But I asked him just to make sure.

"What will become of Nashobah, you mean?"

217

"Well yes, and I suppose this place here'll carry on, I've taken care of that. But the rest, things as they are, I don't know."

Mr. Couper was always attentive to matters of land and had for years maintained his farm under a state law, Chapter 61 A, that reduced taxes for those using the land for agricultural purposes. He had also probably taken care, although I was almost afraid to ask, to make certain that his part of the world would carry on as a farm. As for the other farms, one would be pressed to say.

⸎

But at least things were looking up at the Frost Whitcomb property.

Linda was not the first impediment to the sale of this property, it turned out. The land had been on the market, off and on, for years, but in the past, every buyer had been wary because of the questionable drainage in the area. The current developer had hoped to circumvent this problem by putting in a localized sewage treatment plant that would take care of the waste for the entire development. But he had conveniently overlooked the fact that such plants are controversial and that there was a moratorium on their construction in the town. After he pulled out, the land was back on the market, and suddenly Linda and her group, rather than barriers to sale, became the great hope of the sellers. The owners never were in favor of a big housing development anyway. They even lowered the price when it became apparent that the land might be preserved. What had been a million-dollar sale was now offered at $700,000. It only remained for the Friends of Open Space to convince the town to spend the money.

There are unanticipated economic consequences to breaking up large tracts of land, which, if they ever became common knowledge among town officials, might slow development of single-family houses. It is cheaper for a town to buy land for conservation than to let it become a housing development. Open space does not incur any expenses for a community other than theoretical lost tax revenue. If, on the other hand, a given tract of land is sold for new houses, even though the householders pay taxes, because of the increased costs for schools, roads, fire, police, and the like, the town will not necessarily cover its costs with taxes. This accounting has been known since the 1970s, but so far the message has not reached the consciousness of many town officials, at least not in the towns around Nashobah. So Linda and company were facing a hard sell.

One of the legacies of the Puritan founders was the concept of the town meeting, a more or less sacrosanct institution to which the entire community is invited each year to vote on the business of the town. This eminently democratic system, a sort of rite of the New England spring, can stir the dormant passions of a town whenever there is an issue worth fighting over. Even a single individual citizen can get an article voted on in the meeting if he or she can get thirteen petitioners to sign a request for the article. With this in mind, the Friends of Open Space drew up a petition and began circulating it in the weeks before the Littleton town meeting to get an article on the agenda proposing that the town spend $700,000 to buy the 113-acre tract.

Getting the signatures was no problem; the Friends, by this time, had a major network in the town. Everyone knew someone, who knew someone else, and in a very short time they had more than the necessary signatures and had assured themselves a vote at town meeting. But that was the easy part. Getting the article passed was another matter, partly because of internal manipulations.

There were a few in the community who questioned the idea of spending yet more town funds on conservation lands, although, for the most part, town officials supported the idea, albeit cautiously. But as usual in small-town politics, there was the question of money. Seven hundred thousand dollars was a lot for a small town, and although the money could be paid over a period of time, and although there was a state bond to help small towns purchase open space, the fiscally conservative "bean counters," as Linda called them, were balking and putting up impediments and asking questions. Furthermore, Linda and a few others in her group came to believe that there were certain officials in the town who, although publicly supporting the purchase, were in fact opposed to it and were working behind the scenes to defeat it.

Much chatter in the town. Many late-night meetings, many phone calls and gossipings and accusations. Rumors of alcoholism, of affairs, of secret real estate deals, personal aggrandizement, profit mongering. "The usual," said the Solicitor.

Then came the big night.

The good people of the community began filing into the school gymnasium, where the meetings are held, and by the time the meeting began, the place was filled. The evening droned on with many boring discussions over line items in the town budget. Discussions of sewage, of schools, and of lighting for playing fields.

One of the issues the Friends of Open Space had to deal with was where on the agenda the vote on the land would fall. Early in the meeting, you get a good crowd. But as the evening plods along, the tedium and the discomfort set in and people begin to leave. Late in the night some town meetings have less than half their original participants. There had been charges that the article on the land had been shifted—purposely—to a late position in order to weed out the indifferent voters, most of whom, it was argued, would have supported the purchase.

One of the other issues was that the selectmen, for reasons that were never made entirely clear, at least not in the eyes of the Friends of Open Space, had proposed a purchase of their own. They wanted to buy only one of the two properties, leaving the other up for development. The proposal weakened considerably the position of the Friends.

"Some kind of end run or something," Linda said. "I don't know what the hell they're thinking. Put in a school or something, once you've got the land as open space. We want it for open space, so people can walk there and enjoy flowers and the woods and fields. We don't want a school. Schools are ugly. That's not the point. You've got to breathe."

Fortunately, at the last minute the selectmen decided to withdraw their article.

The meeting moved on slowly, article by article. Fully half, if not more, of the participants had come this night specifically because of the open space vote, and discussion over the other articles was decidedly lackluster. One after another, without prolonged debate, the articles were proposed, voted on, and passed or failed. Then finally around eleven o'clock Linda's hour arrived.

The Friends, armed with maps and charts and financial statements, slowly presented their case. Linda showed slides of the property, spoke of the benefits of preserving open space in general, the lack of new open spaces over the past few years, and the unique quality of this particular site, with its open fields, its rare grassland birds, and its wild, lakeside banks. Then the debate was opened to the floor.

There were many passionate defenders in attendance, and not many opponents, or at least not many who spoke out. One of the finest presentations came from a man who was head of the Massachusetts Association of Conservation Commissions but who happened to live in the town. Given his position, he had seen many such fights across the region, he knew of the difficulties in acquiring open space, and he made it clear that this was a unique opportunity, that open space was disappearing, lot by lot, field by field, across the whole United States. "If not now," he intoned, at the end of his presentation, "if not now, when?"

Thunderous applause from the assembled. Much pounding of the gavel from the moderator.

More speeches, more support. Then a few small whimpering voices questioning the proposal, primarily because of the money, which, no matter how you looked at it, wasn't much–individually.

While these sort of meetings carry on, the officials of the town–the selectmen, the financial officers, the moderator, and the town attorney–sit in the front of the room, at a long table, leaning on their elbows, adjusting their ties or hair, and shuffling papers in front of them. I started to daydream while I was there, and began seeing the officials as seventeenth-century figures, our Puritan forefathers in their black robes and white collars. They looked like alms-board members from a Frans Hals painting, somber-faced, red-eyed, some of them. Of all the speakers in the house, of all the assembled–whether educated, uneducated, blue collar, or white collar–they of the board are the most soporific. None ever speaks with any passion about anything. They are always cautious and rational. None of them breaks into tears at the end of their presentation–as Linda almost did. None of them gets nervous or angry. They are cool, Descartian beings, with nothing exciting to say. But they always get a platform, which is more than you can say for the masses, many of whom had hands waving in a desperate need for attention and were overlooked by the moderator. One of my greatest disappointments was that Farmer Flagg, who had his hand raised, and who always gives a good presentation, was missed. (It is to the credit of this man that his intention that night was to speak up *in favor* of the proposal. "It is the democratic system," he told me later. "It is the rational way to save land. Through purchase by fee simple. Not by petty regulations.")

Toward the end of the debate, one of the selectmen who had questioned the purchase, and had even proposed the alternative purchase, was recognized by the moderator. He started off with a clear announcement of his support for this article and then began droning on about some simple questions he had about it. I lost track early on as to what exactly he was saying. But he carried on, dutifully accounting for this and for that, and questioning whether now was the best time, and whether this or that were true, and so on and so on. The assembled grew restless and began to shift in their seats. He carried on nonetheless. There were murmurings. The people began to grumble, and he carried on. Some began to wonder whether this was per-haps the equivalent of a filibuster. Even those who had no opinion on the proposal were mystified by his long-winded speech. And then he made a faux pas.

"We are all for this, really. We want to go ahead with this and would have gone ahead, had this *impediment* not come up."

The "impediment" he was referring to was the Friends of Open Space.

There arose then from the floor a great, passionate howl. Some angry man behind me jumped to his feet and shouted, "Sit down."

Enough, they said. Much shouting. Stamping of feet, much pounding of the gavel from our ever so moderate moderator.

The selectman grumbled, waved his hand, and took his seat, whereupon he adjusted his tie and patted his hair.

Then came the vote.

The moderator explained that it was necessary to have a two-thirds mar-gin in order for this to pass, and because it was such a close vote, as he deemed (no one else would have said that with all the Sturm und Drang in support of this project), we the people were required to line up and cast bal-lots, rather than hold a hand count, which is the normal means. And then, so Linda claims—I did not hear this—our moderator leaned into the micro-phone and said, "Remember. This vote will affect your taxes."

"Some moderator," Linda said. "I guess we know how *he* voted."

The article passed.

But the victory was not yet theirs.

The state of Massachusetts has a regulation that puts a cap on the amount of money a town can spend each year. If a community wants to spend more than the given amount, it must vote to do so, and since the $700,000 purchase would have exceeded the figure, the town had to vote once more, a week later, to spend the money to buy the land.

Again the network of phones of the Friends of Open Space began to chatter. Again the opponents of the purchase began to talk, and again the voting booths opened.

A few days before the critical vote on the override, there was an article in the local paper stating that the figures announced at the town meeting were wrong and this purchase was going to cost more in taxes than the propo-nents of the purchase claimed. By this time, after a year-and-a-half struggle, and after certain twists and turns in the course of the escapade, Linda Can-tillon was getting paranoid. She had come up against so much opposition to her project, so much backpedaling and stalling and proposals to rethink this whole question and all the usual cautions, that she had come to believe—some say quite justly—that officials in the community were working against her.

"These gorillas," she said, after she read the news story. "Believe me, I *know* these gorillas now. They'll do anything. But this is really too much, feeding false information to the press. It's collusion of the press. What is this, 'Isvestia'? A state-run paper? Give me a break."

It turned out, or so it was said, that the reporter not only did not have his facts right, as reported to him by town officials, he didn't check his sources, and wasn't telling the whole story in any case. The editor, who was more interested in sports news than politics, didn't bother to check either. But the damage was done. There would not be another edition of the paper, correcting the story, until after the vote.

That was on a Thursday. The big vote was Monday night, and for four days straight the Friends of Open Space worked the phones, trying to clear up the misinformation in the news story.

On Monday it rained. By evening it was raining hard, and it was cold. But once more the good people of the town turned out to vote. And once more there was a snag.

Apparently the polls had not opened on time that day. Members of the ever-watchful Friends of Open Space were there at the appointed hour to petition, and working people who showed up to vote before their morning commute couldn't get in and went off to work. The Friends had to call the police, who came and opened the doors to the polls. Then it turned out that the party responsible for opening the gates was also responsible for counting the votes and was known, or at least rumored to be, opposed to the purchase. Small-town politics again. Small-town paranoia rearing its feisty head.

"That did it," said Linda when she learned this. "I'm calling the attorney general's office."

All that day while the voting carried on, calls went back and forth between the Friends and the state attorney general.

None of the Friends believed that the vote count would be fair, and they wanted someone from the attorney general's office to come out and oversee the count.

The group's concerns were heightened when the counter refused to allow the group to watch and made them stand behind a line. But in the end it was another victory. The town voted to spend the money.

It was the final vote, and one would think that Linda and company would be cracking the champagne bottles, but by then they were so tired, so exasperated by it all, they couldn't believe it was true.

They were right.

Three days later news came out that there was a petition circulating to hold a special town meeting to rescind the vote on the open space.

The proposal came from a local man who felt that the town was spending too much money on the open space when there was already a lot of open space in the community. All the realtors in town but one signed his petition; after all, this was a bad precedent, this removal of private, undeveloped land and putting it into common land. Bad for future sales. Bad for the economy. Within a few days, the petitioner got the required number of signatures and a special town meeting was called to rescind the decision to buy the land.

This was a unique case in the annals of town meetings, I was told by the town attorney. Only one or two times in the history of the Commonwealth had town meeting votes been rescinded, but apparently anyone can do it.

Linda was almost broken. I saw her a few days after the signatures were collected and she looked totally worn down, as if she had been crying for a long time.

"I don't know where to turn with this," she said, quietly. "Some gorilla there on some board, some geek is out to crush this thing. We don't know who. We don't know why. Everybody's for this. The people are for it. Why? We don't get it. But I'll tell you one thing," she said, recovering her fire, "if this jerk can rescind a vote, so can we. We're going to be looking at special town meetings up the yazoo if this passes. We'll just go back, get our signatures, and call another town meeting, and then another. We can play this game too, you know what I'm saying?"

I couldn't figure it out either, so I called the Solicitor.

"Look to the deeds," he said. "Somebody's got an interest in some land nearby. It's all money, all greed. Get her to find out who owns the land around the tract, then you'll find the perpetrator. Follow the source, you'll come to the bottom of it, and I guarantee it will be greed. Check our Raging Bull, he's probably behind this."

"She's smarter than you think," I said. "She did that. She knows everything there is to know about her so-called gorillas. There's no profit in it."

"There is," said the Solicitor. "There always is."

He was probably right. but he was wrong about Nashobah.

One night Linda had a call from the town attorney, who was a behind-the-scene supporter of the purchase. The attorney explained that the selectmen had a purchase and sale agreement for the property. If they were to sign that agreement *before* the town meeting, although the meeting would still have to be held, they would, by law, have to withdraw the article to rescind the vote, since they had already abided by the vote of the earlier town meeting and the override vote.

There was to be a selectmen's meeting in a few days, the attorney explained, and if the people showed, en masse, if it were possible to encourage, force, or otherwise convince the selectmen of the importance of signing, they would perhaps sign, and the purchase would go through.

Once more the phone network began to rattle. And once more on a rainy night, the rooms were packed. So many people turned out that there was standing room only and attendees were spilling out into the hallways.

The meeting proceeded, with the usual lineup of good burghers sitting at a long front table facing the crowded room. Much adjustment of neckties, much patting of hair, many throats cleared, and much shuffling of papers. And in the galleries, the restless, land-hungry masses, crying for open space.

I was reminded of a woodcut I had seen in some old history book of the eighteenth-century peasants with pitchforks crowding the court chambers during the French Revolution.

"I guess we should proceed with the business at hand," said the head selectman.

"I guess we should," shouted the crowd in so many words.

There ensued then many comments from both sides, with the division quite clear. A slow cautionary advance on the part of the selectmen, a clamor for action from the people. There was a small matter of detail, a perceived stalling from the first selectman. Linda, who had the floor, began shooting sparks. This had been going on too long, and the will of the people was too clear to wait anymore and Linda had nothing left to lose. She began to say what needed to be said about this whole matter, and the board was not taking it well.

"Linda, maybe you should just sit down and let us speak," one of the selectmen managed to say. "Maybe you're out of order, in fact."

"You sign, I'll sit," she said.

She still looked as if she had been awake for sixty hours.

Yet again the assembled stood, one by one, to express support. No one there was in a mood to hear why it was necessary to proceed with caution and make a careful accounting of this, save one man.

Toward the middle of the meeting, a well-dressed man with a scarf draped over his dark suit was recognized by the moderator.

He began by saying that he supposed he was outnumbered but that he was the man who had started the petition.

One would expect, given the mood, howls of execration at this point. Instead there was a stillness, a polite, somehow ominous silence.

He spoke, mouthing the usual cautionary tales of overspending and how he felt this was too much money to spend on a piece of land that probably couldn't be developed anyway because it was too wet, or had too much ledge, and wasn't there already enough open space in the town? And then he said what I perceived to be the raison d'être for his petition. "Why do we set aside open space where you can't hunt and where you can't have motorized vehicles?"

"I could be proved wrong," he continued. "I just think we should put it to a vote."

"We did vote," someone shouted. "Twice."

The brave gentleman sat down, and a strange silence hung over the meeting, as if to say to the board, "Your move."

"Anyone else have any comments?" the moderator asked.

Another well-dressed man in the front row got up tiredly and announced in a quiet voice that he was head of the Sudbury Valley Trustees, a land trust that is forever involved in issues of this sort.

In a dispassionate manner, he said that what had happened in this town was extraordinary. Rarely, he said, do you see such support and such a willingness on the part of the voters to put money out to save land.

"Win or lose," he said to the selectmen, "these people, these Friends of Open Space, are to be congratulated."

I think it was the convincing argument. The selectmen closed that section of the meeting, and the people filed out quietly and stood in little groups, discussing the outcome.

At their next meeting a week later, the selectmen signed the purchase and sale agreement.

"I don't believe it," Linda said. "I can't believe it. It's been so long, so hard."

I stopped by Bud Flagg's farm stand at Nashobahside after the big decision, partly to sound out his reaction, and partly to see if I could find out his position on the future of open land in the area, namely his land. I suggested, obliquely, that there were government programs available with funding to help preserve farmlands. He assured me that his land, as with Mr. Couper's, was registered with the state to reduce taxes. But his trust only went so far.

"Pretty soon," he said, "things going the way they are, you're going to have to get a federal permit to go to the bathroom."

He fixed me with his blue eyes over his plastic glasses. I could tell from his look where *he* stood on the issue of common land, and asked no more.

Unfortunately, some weeks after this, I saw an ambulance in front of his house, and I later called his nephew, David Flagg, to ask what had happened.

"Off to the hospital, to die," he said. "But they brought him back. He wanted to die at home."

I saw his obituary later that week.

Mr. Couper, who had been lamenting the loss of all his old friends of late, was onto one of the essential problems of the redistribution of land and that is the simple fact of human mortality. How does society go about reorganizing lands after the death of an individual who holds title to the property?

In feudal England, the redistribution was clear, all properties went to the first-born son upon the death of the lord. And upon the death of a tenant, whether villein, cottar, or thane, all lands reverted to the lord, who then redistributed them—usually to the same family. If the tenant died without heirs, the lord redistributed the land among the other tenants. If his children were still living but underage, the lord would assume not only the lands but the children as well. He was responsible for feeding and clothing and educating them until the children were a certain age, at which time the lord would either restore the properties to them or make other arrangements.

This ancient system, along with the problem of the property rights of bastard children, has led to many complex legal and moral battles, and has been the stuff of some of the finest novels in the English language. It also played a part in the settlement of colonial lands. Evicted, propertyless younger children and, later, landless European peasants undertook the dangerous passage to the Americas, drawn by the lure of private ownership. Interestingly enough, having settled on these shores and established themselves as Americans, these families, or their children, or grandchildren, often undertake sentimental journeys of return to their native villages. Some even go back after generations of absence.

Some years ago, since I was in the general area anyway, I decided to visit my own ancestral lands. I am not one of these Americans who is obsessed with family ties, and I only half believed the stories I had heard from other family members of the place—a highland manor house in the small village of Kintore in Scotland, two entrepreneurial cousins who set out in 1722 or thereabouts to seek their fortune, ended up in Virginia, and became tobacco exporters, made money, and built manorial houses of their own at Port Tobacco and joined mainstream American history. I much preferred a story told to me repeatedly as I was growing up by my two older brothers to explain my errant behavior. They claimed that about the time of my appearance in the family, my father, while traveling in southern Europe, had purchased a baby from a band of passing gypsies and brought it home as a gift to my mother.

The town Kintore lay east of Aberdeen in rolling foothills stretching west to the highlands. The manorial estate, known as Thainstone, lay just outside the town, I was told. Kintore was a little gray village so small that it did not even have a bed and breakfast, although I managed to find a place with a woman named Mary, who was a nurse and had horrible medical books from the turn of the century lying about her parlor. My first evening in the town, I ran into the vicar himself, a fine-boned, gray man who walked with a cane and who fixed me with a hawk's eye when I told him why I had come. "Och," he said, "not another Mitchell come to search his roots." It turns out my cousin had been there not three months earlier on the same mission.

The next morning I spent some time poking around the old burying ground looking for family graves. There were many gravestones in this churchyard bearing the name Mitchell, but none had the actual family name, which at that time was Forbes-Mitchell. Because of this I imagined another, not unlikely history for my family. We had perhaps come out of warring highlander stock, or better yet, highwaymen, or footpads, who had been captured on the roads, sent to Newgate to languish until death by hanging in front of a jeering mob. And then on the gallows, this sole progenitor had had his sentence diminished and was transported to Virginia with his highwayman cousin as indentured servants, where these two known criminals made their way in the world by selling drugs (i.e., tobacco).

Just before I left the churchyard, I saw a wall and some stones on the other side, much covered with brambles and nettles. I scaled the wall and dropped down into the yard, and there they were, one after another, the whole family, back to the fifteenth century, interred in this singular space.

After that, in the lingering early summer light, I walked out across the land to watch the skylarks rising and falling against the western horizon, and heard, in the distance, as if to emphasize the essence of this place, the curling whine of the pipes. Back in town I found the source, the local pipe band was marching back and forth in a school yard, practicing for funerals and other happy occasions.

The vicar was there admiring the pipers. "You are welcome to go up tomorrow to Thainstone for tea, if you like," he said. "I've arranged it all for you."

Thainstone was no longer in the family. The last Forbes-Mitchell had died in the late 1940s or '50s and the manor had been taken over by a family named Valentine. The house was built in 1582 under a grant from King James I and was owned in the seventeenth century by a man named Sir Andrew Mitchell. In one of these generations, in order to save the estate, the laird or lord married a Forbes and agreed to double the name to Forbes-Mitchell. In 1722 one of the children, a man named Hugh, left Scotland for the Americas.

Thainstone was a gray-brown eminence set at the top of a sloping hill with a long winding driveway. Beyond the main house, I could see the original manorial lands—pastures, groves of fruit trees, and plowlands, still in production, presumably, since 1582, when the original laird established this place. The house itself was one of those gloomy English stone constructions with a heavy slate roof, a gravel drive, gardens, and in back of the house, as fine a vegetable garden as I had seen anywhere in all of Scotland.

Mrs. Valentine was a gracious woman who served my wife and me tea in a well-appointed library lined with eighteenth-century books. She had, I thought, a lonely, sort of winsome air and told us, sadly, that her husband had recently died and that she was carrying on in the face of it. I gathered her husband had been an older fellow; she herself must have been in her forties. She gave us a quick tour of the house, the old, heavy portraits, more libraries, and stairways, and then, in an upstairs hall, she showed us what seemed to be her pièce de résistance. It was a cheap portrait of a Spanish flamenco dancer in a red dress.

"I do so love this silly little painting," she said. "Just look at her, she's so—so free—so, what should I say, passionate."

It was a horrible picture, all garish reds and flaming backgrounds and hideous golden earrings.

I was more interested in the grounds and the sense of place in this area, of which Madame Valentine knew nothing. We left to poke around the gardens. I have to confess that I stole some radishes and fed upon them in the spot. I thought it might help restore some lost spirit of the place and anchor me in this site. I should also confess that, the day before, I had tried to convince my wife to make love in the graveyard, among the tombstones and the nettles, to see if we could not conceive a child in this ancestral company and thus replenish the blood. Wisely she refused.

I had to wonder though, eyeing this grand estate and imagining how it must have been in the 1720s, if some of my original theories of the Mitchell background might not have been accurate. Why, living here among the nut groves and the gardens, and the outrolling fields below the hill, would anyone want to endure a three-month sea voyage across the North Atlantic and settle in the malarial lowlands of Virginia, where the survival rate of the English was very low. This Hugh Mitchell must have been no Mitchell at all, but the bastard son of Sir Andrew himself, who one day, so I imagine, with his filthy locks parted across his forehead and his broadsword clanking at his waist, leapt over the wall in this same garden while Sir Andrew was taking his morning constitutional. Here, in this place, driving Sir Andrew against this very wall, sword at the old man's throat, he identified himself as his own bastard son and demanded passage to America, which, along with a promissory note, he attained. In short, he was no gentleman farmer, but a remittance man, sent off to America to keep him out of the way.

That is not an unlikely scenario, the colonies provided a convenient dumping ground for social embarrassments. But it turns out it was not quite the case with this family. This Hugh was the second-born son of Sir Andrew, and under the system of primogeniture, which was in practice in Scotland at the time, the manorial estate on the death of Andrew would go in its entirety to Hugh's older brother. So when old Sir Andrew died in the 1720s, Hugh himself was left with nothing. He had no lands, no property anywhere, and perhaps only a little silver. The traditional choices for him before the colonization of America were either the military or the cloth. So this Hugh and his cousin took passage, landed at Port Tobacco, and, according to family historians, began exporting tobacco.

This system of primogeniture, in which the first-born son gets all, was not always fair to the children, but it did make the distribution of lands less of a problem, a very simple case, legally. The custom endured here in the Americas to some extent. But in seventeenth-century New England, in the

wide forests and theoretically unclaimed lands, the egalitarian Puritan traditions transformed the system and came up with another structure, which
endures today. Lands were broken up and distributed, albeit not always
equally, among the children. The result was the creation of smaller and
smaller farms. When the farms became too small to be productive, the
colonists moved outward, away from the village centers.

My first cousin, Jim Mitchell, who had visited Kintore a few months
before I was there, and who documented much of this history of the family
in a book he published, actually may have had, so he claims, a good crack at
regaining the former manorial estate. He met this same Madame Valentine
and came away with a different story.

He too was ushered into the book-lined library, where a service of coffee and scones had been laid out. Madame would be down shortly, he was
informed by a servant. While he waited, my cousin, a single man in his mid-
forties, eyeing the furnishings, began to imagine said madame: a weathered
horsewoman perhaps, hard-boned and dressed in proper tweeds, with
clipped Scottish brogue and a no-nonsense configuration. He was disabused
of this notion with the appearance of the real Mrs. Valentine, who, profess-
ing to have a cold, was attired in a quilted dressing gown, cut low to accen-
tuate a striking cleavage. My cousin couldn't help wondering, in an account
he wrote up of this event, why a woman with a cold would wear such a low-
cut garment in so chill and drafty a place. She was full-figured, with wavy
auburn hair, and had, as he wrote, a peaches and cream complexion. She was
also about his age.

According to the story she told him, her husband was not by any means
dead, as I had been told. She was recently divorced, and he had been sent
packing. Now the place was hers.

They shared their coffee and began to talk about one another. Both had
two daughters away at school, both shared an interest, obviously, in the
place, and not long into the conversation my cousin was informed of an
unexpected intimacy. The estate manager had been making untoward
advances in Mrs. Valentine's direction, now that her husband was gone,
and she did not know what to do about it. Oh my, thought my cousin, how
uncouth.

He was then informed that madame was in possession of a villa on
Minorca, where my cousin, were he so inclined, would be most welcome.
And furthermore, it was ever so lonely here in this empty house, with its
drafty hallways and bedrooms, and the stalking groundskeepers, and the

lecherous estate manager. As madame leaned forward to pour more coffee, her dressing gown slipped ever so slightly across her breast, and my cousin, gentleman that he is, averted his eyes and proceeded to admire further the ancient portraits on the wall and the eighteenth-century book collections.

In short, my cousin, an innocent abroad, slowly understood from the various innuendoes that he was being propositioned. Or so he imagined. His mind began to work. He was single at the time, footloose. Perhaps a visit to the villa at Minorca, a successful pairing. More assignations, and then, finally, marriage. He would recover the ancient ancestral lands. He would assume the title of Sir James Andrew Valentine-Mitchell, now knighted for his efforts. He would follow tradition and save the place by uniting with a powerful female line.

But then what would he do in this cold, rainy, windswept place? Compose his memoirs? He was too young. Produce fruit wines, perhaps, from the multifarious raspberries that grew here? (This was a recurring dream of his; later in life he created Sakonet Vineyards, in Rhode Island.) But not here, too cold. He would take over the single-malt whiskey business that was already in operation there. He would write. He would go to sea (another dream of his also realized). After successful voyages he would return to doze by the coal grate over his whiskey.

But none of these fantasies caught on, and in the end, Sir James, the manqué, arose from his couch and shook the hand of Mrs. Valentine, and thanked her for her hospitality.

There was not a word of this when I met Mrs. Valentine. She seemed a perfectly charming, upright person, a horsewoman perhaps. The only crack in the mirror was the upstairs portrait of the Spanish dancer in her low-cut red dress.

<hr />

One of the other people I had hoped to talk to about this problem of inherited land was the old dairy farmer known as Chicken John, whose family was one of the original purchasers of Indian lands in the Nashobah area.

One rainy summer day I went over to his old farmhouse to ask if I could come by and talk to him sometime.

"What d'you want talk about?" he said abruptly.

"Land," I said. "Land and old man Morrison."

"Well, if you want to talk about Jack Morrison, you go over and talk to Vint Couper, he's the only one in this town could stand the man."

"Well land, then. I just want to find out more about his land."

"I can't talk about land."

I asked in so many words—I was trying hard for an entry here—what he would like to talk about.

"Cows," he said.

"Well that's good. I like cows."

"What do you know about cows?" he asked.

"A little. My cousin had cows."

"Well then, do you know the difference between a four teater and a three teater?"

I winged it and got the right answer.

"All right, which brings more on market, three or four?"

"Four."

"Wrong."

"Three."

"Wrong. Don't make any difference whatsoever, three teater'll give just as much milk."

Somehow I passed muster and was invited back to talk more about cows.

Usually Chicken John, who was ninety-two or -three at the time, could be found sitting in an easy chair just inside his back porch. But when I went back for my cow discussion, I could see through the window that he was not in his chair. The lights were on, though, and the door was open, so I walked in and called. No answer save for a slow, steady drip from a faucet in the kitchen sink, which, water conservationist that I fancy myself to be, I turned off. Then, finding no one about, I left.

It occurred to me that the old man was sharper than I thought, and I had said Friday, not Thursday. So I went back the next day at eleven.

The same scene greeted me, the empty chair, the light burning, doors wide open, and Chicken John nowhere to be seen.

One of the best places in town to collect gossip is the assessor's office, so I went over to the municipal building and began asking questions.

"He's in the hospital," I was told. "Checked himself in with the flu."

I saw his obituary the following week in the town paper, a long praiseworthy essay. He was one of the more beloved old farmers in a town that has had no shortage of beloved old yeomen.

"It is," said Mr. Couper, "one of the scourges of old age. You live too long you lose all your friends."

Mr. Couper and his cousin Chicken John were not close friends when they were growing up, but time threw them together and in the last decades they had become good companions to one another. John Adams Kimball became the proofreader and copy editor for the vast tome entitled "Memories from the Snake Pit," which Mr. Couper was in the process of writing, and the two of them would often be seen together around the town. On one occasion, Mr. Thomas Todd, a well-known Boston printer, invited the two of them to come for lunch at his house, not far from the Morrison land at Nashobah. They were instructed to come at twelve o'clock, and Mrs. Todd would entertain them until Thomas returned from another appointment.

The two old men proceeded along Nagog Hill Road in Mr. Couper's white 1979 Buick, turned as instructed at Fort Pond Road, and followed the street numbers to the house. It was a well-appointed dwelling with much landscaping and a smooth-paved drive. The garage door was open and there was no car around, and in fact no one answered the doorbell.

"Gone to town for food," Mr. Couper said to John Adams Kimball. "Probably be back in a minute."

The door was open, so the two old men went inside. These two were self-described dirt farmers, and although they had gussied themselves up for the visit, they were in an alien environment among the Chinese antiques, the Oriental rugs, and the clean white kitchen. But the place was comfortable, and inasmuch as Mr. Couper had had a long morning in the greenhouse, he lay down upon the couch, and took a nap.

Chicken John was less comfortable.

"I never did go to an Ivy League college," he said later, by way of explanation, a tidbit of information that explained nothing, in fact.

Mr. Couper's nap was interrupted by the postman, who arrived at the door and dumped the mail in through a slot.

Chicken John eyed the mail and remained seated.

"Wasn't any of my business what was in Mr. Todd's mail, was it?"

Not so with Mr. Couper. He rose from his resting place and began to sort through the mail.

"Nothing here for Tom," he said. "Most of this is for someone named Marjory."

He went back and sat down. They stared out the window at the grass and the rhododendrons and the birds at the bird feeder.

"You say Marjory?" Chicken John asked.

"Marjory."

"No Marjory here. Mrs. Todd's name is Barbara, I believe."

"Barbara?" said Mr. Couper.

"Barbara," said Chicken John.

Mr. Couper unrolled himself once more from the couch, picked up the mail, and stared at it for a long time.

"John," he said, still looking at the address. "I do believe we're in the wrong house."

Chapter Sixteen

So God created man in his own image, in the image of God created he him: male and female created he them.

And God blessed them, and God said unto them, Be fruitful and multiply, and replenish the earth, and subdue it: and have dominion over the fish of the sea, and over the fowl of the air, and over every living thing that moveth upon the earth.

—Genesis 1:27-28, King James Version, 1611

The Intelligence
of Salamanders

After a light snow that winter, I was walking the northern boundary of the tract and got stuck in a swamp. There was a sharp little hill just west of Nagog Pond with a flat wetland at the bottom, covered with ice and snow. It had been relatively cold for a few days, so I assumed that I could get across the swamp without having to tangle myself in the thickets that surrounded it, but halfway over, the ice gave way and I went up to my ankles in mud. The smell of spring gushed up with the wet earth and water into the brittle air. I climbed up onto the ridge through the thickets, found a dry sunny rock, and sat down to empty my boots.

Such is the business of the surveyor. Samuel Danforth and company, who walked this same line—presumably—in 1686, would have had similar problems. Since they were running out straight lines, with little regard to the nature of the landscape, they would have had to cross through swamps and dense thickets of brush with mosquitoes, or as one of his compatriots wrote, the dangers associated with animals such as lions, lynx, ounces, wolves, and bears.

Actually, in Danforth's time, this whole site might have been more open than it is today. The Indians of the period were skilled game managers, and it was customary, in some areas at least, to burn over the woodland to keep down the underbrush. This had two benefits for the Indians; it encouraged the growth of blueberries, which were a favorite ingredient of the stew of game, berries, and roots that the Indians would feed on day after day, and the blueberries would attract also bears and deer, which the Indians would then hunt.

Even today there are pitch pines interspersed with the oaks and hickories of the Nashobah tract, especially in the Sarah Doublet Forest, at the center of the tract. These pines generally grow in areas that have been burned over. But these particular trees probably sprang up after a fire that swept over this area in the nineteenth century. And in any case, if Nashobah were, as some believe, an important spiritual area for these people, they may not have dared hunt there for fear of catching some greater game, the spirit bear, perhaps. They may not even have farmed here, according to some historians; and some anthropologists suggest that they might not even have frequented the area. For example, Indians would often avoid mountain peaks such as Katahdin, where they believed important deities lived.

From the swamp I crossed over the four corners at the Morrison orchard, and then cut into the deep woods on the north side of the pond, and, weaving between house lots, came into an area of light woods cut through with the ubiquitous stone walls. At the northern point, near a site known as Cobb's Pond, I turned west. Within a quarter of a mile or so, still zigzagging to avoid people's yards, I came to a new development, skirted it, too, and then found myself in a pine woods. In front of me, there was a stone pile about three feet high, and oblong. Ten yards away I saw another, this one perfectly round, and then another. There were more scattered all through the woods.

I knew of these "monuments," as they were called locally. They had been discovered by surveyors who were laying out the land in preparation for the new development, and a certain faction in the community had determined, or proposed at least, that these were Indian burial sites, perhaps one of the "islands" cited in the nineteenth-century literature where the remnant band of Indians would periodically come to camp. The developer proceeded in spite of the grave sites, and more than half the stone piles were destroyed. But a few remain, some of them at the very boundaries of the backyards of the new residents.

I started following these stone piles, one to the other, and was soon so engrossed in my quest that I actually stumbled into someone's yard. There was a good yeoman of the village standing there with a splitting maul in his hand, dressed in flannel and jeans.

"Can I help you?" he said, suspiciously.

I couldn't blame him for his suspicions. I was actually hot from scrambling through the brush; my jacket was open, my shoes were blackened with mud, and little fringes of ice and mud were clinging to my cuffs. I apologized and explained my mission, and he grew friendly, even enthusiastic, and claimed to know all about the monuments.

"Let me show you something," he offered.

I went with him to the north side of his suburban yard to a break in the stone wall in that section. There was a flat stone in the break, somewhat smoothed over.

"See this site?" he said. "This is where the chief used to sit. He would watch the sun go down from here on the summer solstice. They used to sacrifice girls over at another place near here. You can see the rock where they stretched their victims, there's like a flat place there, with a shallow bowl to catch the blood."

"Why would they want to catch the blood?" I asked.

"You know, it was a sacrifice. They would drink it. The chief would drink it first, then pass it around. They did that before they attacked."

I wanted to know who told him all this.

"The guy who built these houses. He knows all about Indians. Him and the developer. The developer's like an archaeologist. He's got all these arrowheads he found here. Or someplace. They used to have dances here, get really worked up, then they would attack the settlers. You can't blame them, I'd attack too if some guy tried to take my land."

Armed with this knowledge, heart in hand, lest any of these bloodthirsty natives return from the grave to sacrifice me upon the rock of death, I pushed on.

He was right about the dances, though.

One of the most thorough accountings of the local tribal customs in time of war comes from the narratives of captivity by Mary Rowlandson, who was captured at Lancaster and traveled with the Indians until Sarah's husband Tom negotiated her release. She described a ritual dance in which the Indians performed a sort of mimetic drama of a battle, with a warrior standing in the center of a circle with a rifle, surrounded by ritualistic attackers. The raiders evicted the armed man from his stronghold. Having danced the story, they went out the next day and attacked the town of Sudbury.

I asked the Solicitor once what would happen if the descendants of these same Indians, the Muhucanuh Confederation, put the case that inasmuch as the Pawtucket Indians went about their business by consensus, and inasmuch as Sarah Doublet *personally* deeded over the property to the Jones family, even though there were other associates of her people living in the area at the time, could they not argue that Sarah's transfer was illegal and that the land still belonged to the Nipmuck or the Pawtucket people?

"Case denied," said Mr. Solicitor. "Were the Nashobah people a tribe?"

"No, they were a group, probably Pawtuckets and Nipmucks who lived together and were referred to as the Nashobah people."

"Your case denied. They have to be a tribe."

"Say they were Pawtuckets."

"Recognized tribe?"

"Yes."

"How long, how many? Where do they live? I don't know any. Never heard of them."

"You're an English fellow. You don't know anything about Indians."

"I'm the federal government, and I do not recognize these Pawtucket as a tribe. Case denied."

He downs his Guinness and winks.

"Here's another then: Put the case," I said to his Honor, "that in this place called Nashobah, there are certain spirits walking abroad, and that these spirits have been detected by no less than three different cultures: the Indians, who thought the woods were alive with living entities that they could actually see and for whom they even had names; the Puritans, who agreed with the Indians that there were spirits in the forest, evil ones; and finally, certain parties within our own modern society who believe that there are gnomes and ghosts around and about in these hills which must be propitiated if the world is to continue. Given the presence of these beings, why could you not argue that these entities require this site in order to exist, and if it is altered, if it is developed, they will die, they will disappear forever. These lands, where these spirits dwell, are sacred to two different religions, the land itself is part of their sacrament. If you destroy the place, you destroy the object of their worship, you deny them their religion. We currently believe in freedom of religion in this country. Is it legal to destroy the artifacts of religion?"

"Denied for lack of evidence."

"I've got *prima facie* evidence here. I see the evidence of their worship."

"What evidence?"

"Brush donations, sacred arrangements of stones, placement of pine cones in strategic sites."

He merely looked at me this time and rolled his eyes.

"What about Indian sacred lands in the West? Those have been saved because they were sacred."

"Cite the case."

"Can't."

"Denied."

"Well, I know of one."

"Cite."

"It was a timber operation on a sacred mountain in Arizona. The property holder leased the land out to a logging company. Indians protested, claiming that this mountain was an integral part of their religious beliefs. It was the site from which their people were created. Court upheld the Indians' position."

"What court? I never heard of this. Who owned the land?"

"It was National Forest, the federal government."

"Doesn't count. That's not a private property issue, that's a government decision."

"All right, how about this: The greater amount of living matter in a certain tract of land consists of salamanders. If you alter this property by constructing houses here, you effectively destroy the biomass in that site. Do not these entities have rights? And since they cannot protest, can we not enter, what do you call it, an amicus curiae brief on their behalf, or however you phrase it?"

"No. Salamanders do not as yet have standing."

"What if it were dogs inhabiting a place, can we destroy them?"

"No. Dogs have rights, certain rights. You could make a case not to destroy them, but dogs don't need that particular piece of land to survive."

"Why not salamanders though? Why is a salamander lower than a dog?"

"Do salamanders have a conscience?"

"That is not for us to know."

"Do they have consciousness of themselves, a concept that I am a salamander, or at least I am me, whoever I am."

"We don't know yet what salamanders think. They could be very intelligent."

"Are they those things that live underground for eleven months of the year and come up on rainy spring nights, mate, and then go back underground?"

"Yes."

"Not intelligent," said the Solicitor. "You're on the wrong track here. You've got the whole history of Western thought pitted against you."

"What about Eastern thought?"

"West wrote the law."

So I told him how I had walked out across the Beaver Brook site that was to be developed. In the course of a few minutes, I spotted there a leopard frog, a species that may or may not be endangered, but which is quite definitely disappearing from the grassy fields of New England. I saw a red-bellied snake, which is rare; a garter snake, which is not rare; a bluebird which is a species returning only by the grace of certain energetic people in society who are working hard to reestablish this beautiful bird; and, finally, I saw a red fox. I said I had also seen or heard in this area barred owls calling from the wooded lowlands along the Beaver Brook; I had seen woodcocks there, I knew that endangered blue-spotted salamanders were there, and wood turtles had been seen along the banks of the Beaver Brook and also box turtles, both of them threatened species in the state of Massachusetts. What right, I asked, by what logic do we eliminate these living things in order to construct sixty dwelling places for a total of perhaps two hundred human beings, whereas we will eliminate in the process a whole biological network of mammals, birds, reptiles, amphibians, worms, and as-yet-unidentified one-celled plants and animals that, for all we know, contain the keys to the secrets of the biological universe, not to mention the cure for all human diseases.

"How can we presume to destroy this?" I asked him. "This is hubris, this is arrogance of the worst sort, don't you think?"

"Who owns the land?" he asked.

"A farmer whose taxes have grown so large, even as his income declined, that he was forced to sell to a developer."

"Court recognizes the human owner, not the rights of salamanders. This salamander business. Don't start there. No one even knows what they are. Are there wetlands there?"

"Yes."

"Are they mapped? Delineated?"

"Yes."

"Are they to be filled?"

"No, the development has been restructured to avoid the wetland areas and a pond where the salamanders breed."

The Solicitor shrugged. "*Les jeux sont faits, mon vieux,*" he said. "This developer seems to know the regulations. He abides by the law, he proceeds by stages, and in the process he recreates a world. This is legal."

"So then just because you have a wetland doesn't mean you can't develop?" I asked. "Doesn't mean the government is taking your land?"

"The law never did say that anyway. You said it."

"I didn't say that. They say that."

"They who, they the people?"

"Yes. 'We the people . . . We hold these truths to be self evident, that all men are created equal . . .'"

"I know where you're headed. . . . 'nor shall private property be taken for public use without just compensation . . .'"

"Whatever that means."

"Listen, do you know what I've gone and done?" he said.

"What? What have you done now?"

"I've become an American citizen. Took the oath three weeks ago."

I was actually shocked by this. It seemed so extreme.

"Why on earth would you do such a thing?" I asked.

"I don't know really, it seemed only right, been here thirty years now. I don't know actually."

For the first time since I met him, he seemed not to have an answer.

In 1995 members of the New View co-housing group found a new piece of property about two miles south of Nashobah, next to a piece of farmland and within a few hundred yards of one of the most popular greengrocers in the area, a sort of glorified farm stand with a European flair. The land was part of a former nineteenth-century farm, which had been abandoned for years and had, as has much of the open land in the Northeast, grown up to woodlands. There was a stream nearby, there was a section of sharp little hills, and, surrounding the tract, there were open fields, some of them cultivated, some old pasture. Beyond the pastures there was another tract of low-lying wet woods that had been set aside by the town as public open space. The planner for the co-housing group saw a way of fitting the new houses in among the hills and designed a plan to save and retrofit an existing old house on the property and preserve the fine stone gate and the winding driveway that led to the main house.

The little brook was part of the same watershed that the Jones cousins from Concord had used to power their gristmill in the late seventeenth century. The old Jones Tavern, where Sarah spent her last days, was two miles to the south.

Just down the street, in a small mill house, lived an eighty-two-year-old woman with a sharp mind named Mrs. Smith, whose daughter cleaned house for the Morrisons. The old lady had grown up in the house before the Morrisons bought it and had many memories of the place in the time when the property was a dairy farm. She told me once that her brothers used to bring in old things from the high pastures behind the house and the shores of the pond.

What kind of old things, I wanted to know.

"I don't know, old things, stones like. Indian things. Arrowheads. Fishing sinkers."

She also remembered the night back in the 1920s when Virginia Mills drowned. She was the one who had told me the story of the flashlight gleaming beneath the waters.

"I was afraid to go down there for years," she said. "But I remember the place. It was spooky in that cove. That dark island where the beautiful girl on the horse used to live. And all those stories."

And what stories would those be, I wanted to know.

"You know, the stories of ghosts, and the Indians coming back, and then poor Virginia, drowning and all. It was scary. I was but a young girl then. Everything scared me."

Up the street from Mrs. Smith's, after many a setback, and many a meeting with planning boards and health boards, and building inspectors, ground was finally broken for the New View co-housing development.

Things had not worked out the way everyone expected in the beginning. The idea was to create a development that was egalitarian and would include a variety of race, income, and age groups. They wanted mainly to recreate a mini city in the community, all squashed into one site and living there in harmony. They wanted to make a model for the future. But as is their wont, economic realities had intruded.

For one thing, the group found it was hard to recruit African-Americans, or any other ethnic group, to the idea. These people either wanted to be with their own kind or wanted to realize the American dream of a single-family home with a green lawn. Older people, who theoretically should have welcomed the project because of the low cost of the housing and the instant, mixed community, were also reluctant to sign on. Part of the problem was money. What started out as a cheap housing alternative—the houses were originally intended to be small—began to get out of hand, as some of the members of the community decided to build larger structures. Furthermore, the cost of planning and site preparation, and the increase in the cost of building, drove the economics up far beyond what the original plans had called for. The group scaled back the size of the development, made some alterations, faced some realities, and forged on. They had finally found a good site, open spaces nearby, a train nearby, a highway, shopping, and the draw of the open-air market, and if the original ideals were not entirely met, the group was pragmatic enough to shift its focus slightly. Besides, planning for this development had started seven years earlier. It was about time.

All that spring the ground work proceeded, creating, as these things do, a hideous mess of the land. The property had had a sort of run-down, forgotten aura about it before the development came. The old stone gatepost canted over to one side, the gravel driveway curved up to the old federalist farmhouse, which itself was slightly down at the heels, and there was a vaguely sinister quality to the place, in an entertaining sort of way. I wouldn't have minded living in the old house myself, and ending my days rocking on the decrepit porch, eyeing the curious trespassers who would sometimes venture up the bramble-strewn path to the house.

In the brave world of New View, all this was to be cleaned up. The old house was restored and divided into two apartments. Smaller new houses were sprinkled around it, nestled into hollows or perched on hills, and the whole of it fitted in among the standing white pines. The builder—at the instruction of the group—did his best to preserve what was worth preserving of the native vegetation and the landscape.

When the job was completed some twelve months later, there were a few cynics among observers of the project who saw nothing more in the site than yet another development, this one less grand perhaps and without lawns, but a new development nonetheless.

It would indeed take something of an eye to see what this place would become, not what it was in its raw young state. Landscape takes time to create itself, and I believed that once the vegetation filled in, the look of the place would improve. The houses were scattered around two small greens, interconnected with paved walkways and with vehicles kept away from the dwellings in separate parking areas. Sitting on the porch of one house, overlooking the green, you would see only the other houses and the vegetation on the other side of the common—no cars, no traffic at all, save the to and fro of skating children.

Through Sacha and June, I got to know a few of the people there, and would sometimes go to the cookouts and events that the group seemed to have on a nearly continuous basis. Lacking the common house, which was put off for a couple of years because of expense, the group began having regular Friday-night dinners on a circulating basis. The diversity of this group was not at all what the original planners had hoped for. Most of the families were liberal-minded, even left-leaning, mid-level workers in the small local computer companies. They were for the most part white, middle-class, well-educated, with one or two older couples and a few single women, some with children. No Latinos, no African-Americans, no Asians. The only apparent anomaly was that a large percentage of the families were mixed marriages of Jews and gentiles—or at least former Jews and former Christians; many in this community had no standard religious affiliations or would identify themselves as pagans.

The venture in co-housing was not a great money-saving move as things turned out. Sacha and June found themselves working long hours. Sacha expanded his chiropractic work, and June had to give up a free-lance writing career to take a job with a computer company. They would come home late Friday night from work, often too late for the regular Friday-night community dinners, and crash on their couch and watch television. Sacha had to work Saturdays, and June spent much of her weekend catching up on lost sleep or running errands.

"We're not Native Americans here, if that's what you're hoping," Sacha said once. "We're just good American bourgeoisie, trying to get by."

That may be a bit of an exaggeration. Their small attached house is a veritable museum of eclectic music and literature. The books, which cover one whole wall and are overtaking an extra room, range from Sufi mysticism, to Ezra Pound, to medieval literature and the psychology of religion. The music is world music—Moroccan Nafa, Senegalese liturgy, Mozart, Gregorian chant, and classical north Indian ragas, to name a few. At night, Sacha reports, they often *read*.

That winter the waters of Nagog Pond remained open through December, with only a little skim ice at the edges. Then shortly after, with no snow in the offing, a relentless cold wave settled in, bringing with it temperatures that hovered just above zero even in the day. The result was that all the ponds in the region froze into that smooth black ice so beloved by skaters and ice boaters. Suddenly, on the little lakes and ponds around Nashobah, people emerged from their winter hibernation on Sunday afternoons and, bundled in scarves and wool pants and bright winter shells, skated around in circles in the little coves in the bright air. The happy few even descended upon Nagog and made little curving circles around the island off Morrison's property. They were like the seabirds that gather there even after the ice has formed.

One night during this period, another rare event took place, an absolutely glass-clear full moon. I thought this might be a good time to skate out to the Isle of the Dead to retrieve the compass I had left there back in September, and so on the selected night I hid my car away from the pond, hiked down to the shore, and changed into my skates. The sharp runnels of noise ran along beneath me as I skated toward the island, but halfway there I decided things were going too well to just pick up my compass and return, and so I headed away from the island over to a cove on the southwest side, behind the peninsula where the summer house used to be located and where I had surprised the teenagers the summer before. Close to the shore, as I approached, I could see a shape, like a small collection of rocks, but as I came nearer there was a flurry of wings and the shape materialized into a small flock of birds, probably herring gulls, that wheeled off into the black sky to the west.

I turned around and headed for the island, and after skimming along with smooth, effortless strokes soon came to the shore. I had left the compass on a rock that protruded into the water, so I didn't have to cross through the woods in my skates to get to the site. I could even see it there in the moon-light, where I had left it, four months earlier. The Isle of the Dead is not a commonly visited part of the world, thanks to the trespassing laws of the Concord Water Department. I retrieved the compass and skated off a few yards and then lay down on the ice and stared at the full moon for a while.

The stillness of Nashobah descended. I thought I saw something cross the moon, a flitting shadow, then another, and then a few minutes later another. I sat up. White forms were criss-crossing the pond silently in a slow circle—gulls again, I supposed. Then out of the stillness, I heard a loud crack from the black interior of the island. Half in sport, a sort of semi-false alarm, I leaped to my feet and raced off for the middle of the pond, imagining myself pursued by Hobamacho, the hideous wood demon that lived in the forests of New England before he was banished to the underworld where all such demons dwell. Like a fleeing sinner out of Dante's frozen ninth circle, I dashed away from the Isle of the Dead and skated wildly toward the opposite shore, heading east across the lake to the friendly, well-lit banks of the Great Road. I was in a self-induced panic and seized with some kind of supernatural energy, and I might have continued on to the eastern bank except that, halfway over, I skated into the midst of a frantic beating of wings and sharp cries. Large ghostly forms rose all around me, circling and squawking, and beat away for the opposite shores in all directions. It took me a minute to realize I had skated into another flock of resting gulls, and I sprawled once more on the ice, breathless, and lay spread-eagled there.

Not to be dispersed, the forms once more wheeled against the moon, two or three, then ten or twelve, and soon, the whole crying flock was rising and falling over me, as if I had sprawled in the one place on earth where they had to be.

These gulls present a strange anomaly. I am told by my bird associates that gulls seek out freshwater near the coast in order to bathe. But why they should come all the way inland to Nagog, a distance of some thirty miles, to do so is inexplicable. Furthermore, having flown all that distance, why here? Why at this singular pond? There is an abundance of small freshwater lakes in the area, two of them larger than Nagog, and save for a little development along the eastern banks, both of them wooded and wild. Yet at any time of the day or night, in any season, you will see gulls here. Sometimes from as much as a mile away you can locate Nagog by the white cloud of seabirds that hangs over the place. No one has been able to explain the reason in a rational way. There is another explanation, though.

I was telling my friend Mara about the phenomenon one afternoon as we watched an island-like collection of seabirds floating off the Isle of the Dead, preening and crying and diving in the flutter of bright waters. She knew the history of the place, the story of the Christian Indian families who settled here, believing in salvation, only to be rounded up and hauled off to the alien environment of the cold sea-bound shores of Deer Island in Boston Harbor.

"It seems obvious to me," she said, staring out toward the gathered flocks. "Cultural memory. They've come back."

"Who has come back?" I asked.

"Sarah and her people."

You could understand why this woman and the Solicitor were not able to stay married.

The wheeling birds above me began to thin, and I realized that they were actually settling around me, at a distance, but resettling nonetheless, so I lay still, hoping they would come closer, but after a few minutes I began to get a chill and rose up again. They rose too, silently this time, with less alarm and began their circle. By the time I had reached the shore, I could see, in the light of the full moon, that they had landed once more and were preening and fluffing their feathers and crying out from time to time.

On shore, the night woods were black and still and ghostly. No rustle of mice, no crunch of demon footsteps, no creak of branches. A vast, huge quiet, large as time, settled in. If I had not been by this time shivering with cold, I would have stayed there and waited for something to happen.

Chapter Seventeen

No person shall . . . be deprived of life, liberty, or property without the due process of law; nor shall private property be taken for public use, without just compensation.

—Fifth Amendment to the Constitution of the United States

The Landing

There is a small private clearing below a wooded hill on the west side of Long Lake, opposite the Frost Whitcomb property. The tiny sand beach there was a popular gathering spot for the members of the Nashobah jester chorus. On hot days they would sometimes collect to loll in the waters and gossip about one another. I used to take my children to the place all the time to swim and look for snakes and frogs, and there was always a fine school of sunfish off the west bank that would become very tame by late summer and nibble at your legs if you stood still. My kids spent hours trying to catch them with their bare hands. The landing was private property, in theory at least. It was owned by one of the older families in the town, and to get there you had to drive up a narrow wooded road that had a rusting old No Trespassing sign at the beginning. The owners were lax about who used the little beach other than on certain days of the year when they would hold family reunions there. Over the years, even though none of us owned the property, those of us who were using it on a regular basis became highly proprietary of the little landing.

Fishermen would commonly pull up on the shore at the spot and, for some perverse, atavistic reason, feel that it was necessary to leave little reminders of their visits in the form of a beer can or two, or those hideous little bubble-pack packages that hold hooks and other fishing paraphernalia. Furthermore, the landing was sometimes plagued by the dreaded teenagers who would occasionally make night forays to the place and litter the beach and the surrounding woods. We regulars would find ourselves cleaning up the site, raking the beach, and collecting the beer cans.

This proprietary sensibility led to a number of encounters over the years as we attempted to hold at bay those alien trespassers we would discover there, either swimming or beaching their boats for a few minutes. This was a delicate situation, legally, since we, the archprotectors of the place, had no more rights there than the other transgressors other than the fact that we were all friends of the owner. Our usual approach, or at least mine, was to ask if the invaders were friends of the owner. If they were, we enlisted them as fellow guardians. If they weren't, we felt obliged to explain that the owner did not mind people coming to the site but that he despised litterers, and that he had been having a problem recently with marauding teenagers and filthy fishermen and even with trail-bike types who would roar through the woods from time to time and leave ruts in his sandy beach. None of this was necessarily untrue—all these dark events had indeed taken place at least once over the ten- or fifteen-year period that we used the property. But it certainly was not true all of the time, or even most of the time. Generally, the landing was one of the more peaceful places in the whole region. In fact I often found myself musing on the fact that many in this community would get in their cars, hitch up a powerful motorboat, drive two or three hours in hot, weekend traffic to a cottage on one of the larger New Hampshire lakes, there to sit cheek by jowl with other cabin dwellers, while out on the lake the noisy throngs of motorboats and jet skis screamed across the waters from morning until dark. Here at Long Lake, by the imposition of a simple town regulation limiting horsepower on boat engines, the general public eschewed the place, even though, by state law, they had legal access at an official town boat landing at the northeastern end.

Most of the encounters I had at the landing were civil. After I would explain the situation—rationally—most people would understand. Many even launched into sympathetic diatribes against litterers. (Why I was never able to catch the piggy fishermen, I do not know.) But one or two felt that because the landing was not posted from the water's edge it was their right to be there. One man in particular grew instantly outraged when I presumed to ask if he knew the owner.

"No," he said aggressively, "I don't know him personally. But I know of him. And if he didn't want people coming here then it's his duty to post this land. I don't see any signs here, buster, so I'm swimming here, all right?"

"All right," I said. "Just watch out for those jellyfish."

Long Lake is fortunate to have a unique species of freshwater jellyfish, known scientifically as *Craspedacusta sowerby,* that periodically goes through extreme population explosions in late summer. The jellyfish are small, dime sized, and perfectly harmless, but once you learn to spot them you see them everywhere, you cannot swim without touching them, and if you are not familiar with them, or if you have an aversion to jellyfish, they can be disconcerting. My newfound enemy looked down into the waters.

"Watch out," I said, "there's one now."

He spotted it and said nothing, but backed out of the water carefully.

"Did it get you?" I asked sympathetically.

"Nah," he said. "If it's not one goddam thing, it's another."

Once at this landing I even met my nemesis. One hot summer day I saw a kayak pulled up onto the beach and a man and a woman swimming far offshore. When they came back, I went through the usual discussion about the rights of use of the place. The man, who was a healthy blond fellow with blue eyes and a friendly, direct manner, said he understood and that although he didn't know him personally he had once spoken with the owner. It turned out the kayaker was a developer who would search out large tracts of unbuilt land such as the forest above the landing and seek to buy the properties to develop them. He had called the owner of the landing and the woods with an offer and received a resounding no. (Much to my relief, of course, but I said nothing.)

Said developer was not an entirely popular man around town, he indicated, and in fact was even at the time involved in a fight over a development on the north side of the town, about a mile east of Nashobah. He claimed he was a conservationist at heart, but that people need to live somewhere, and since that was the case, why not do it right. I told him I couldn't agree more and launched into my vision for the town.

There were, as he well knew, still plenty of open fields and forests around Nashobah, and the old commercial center of the town of Littleton had been, as early as the 1950s, destroyed by its own townspeople. The planning board, or whatever committee decides these things, had seen fit to place gas stations next to the old churches, had destroyed a hayfield and a dairy farm near the center of town to put in a huge parking lot with a supermarket and a couple of shops, thereby driving out of business a perfectly good general store with a wooden front porch and fruit baskets and vegetable bins. Same thing happened to an old-fashioned gas station and a hardware store. By the middle of the 1970s the center was nearly obliterated. 257

But at least most of the changes had been centralized. Why not, I asked the developer, put a permanent moratorium on all development of lands outside the village center and then break all the old zoning codes and allow apartments and condos, shops, and businesses and offices to all mix together in the center. Keep what was left of the town green as a park and garden, divert traffic around this center, with parking on the outside, and turn the core of the town over to pedestrians.

"What do you mean?" he asked. "Apartments in the middle of town? Like people would live next to the supermarket?"

"Yes, even above the supermarket, above the streets, next to the offices and shops. You want to go to work, you go down the street to the office. Stop at a café on the way. Go home for lunch if you like. Read the paper in the park. No commute. You wouldn't have to sit in traffic for two hours each day."

"Nah," he said. "That's crazy. People don't want that."

His woman friend chimed in at this outrageous idea.

"What are you saying anyway, like people would live in high-rises right in the middle of the town?"

"Maybe. I don't know. If necessary."

She shrugged.

"That'd never get through," the developer said. "Town would never let you do that. And anyway, I can tell you people don't want that. They all want their yard. Big house, a lot of bathrooms, modern kitchen, big green lawns. That's what sells. Believe me. It's my business. I know what sells."

Maybe he was right. But who would know, since current development patterns and zoning codes have not given the population many choices.

Places such as this landing, when they are open to the public, offer some of the last expressions of the common in America. The town beach at Long Lake, which is at the eastern end of the lake, opposite the fiercely defended landing, is a case in point. During hot days in summer, the place is a typical small-town beach with many screaming children, ever-watchful lifeguards, a few retired or out-of-work people who sunbathe or read, leaning against a whitewashed wall that separates the beach from the parking lot. It is after hours, though, at evening, when the place becomes more like a common.

For those of a poetic frame of mind, the beach is an excellent place to watch the sun go down, especially in the last hot days of early autumn. There are always a few lingering swimmers in the water at this time, fatigued workers sometimes come there, either to take an after-work dip or simply to watch the sun go down, and the lucky few who live in the former summer houses around the beach bring their dogs down for a swim or send their children out to dig in the sand–unsupervised.

There are a few regulars who greet each other, informally, and there are a few close friends who meet each other there, and I know of at least one love affair that blossomed on the beach. On hot days, people linger long after the sun has set to catch the cooling breezes off the lake, and there is a sense of communion that one does not witness at the local supermarket, or at the new drugstore in town, or at the ungracious, unforgiving windy parking lots that replaced the old town in the 1960s.

Nevertheless, the old model is here. In spite of the isolated private manor houses of the American suburbs, in spite of great green lawns, and television, and the isolation of the Internet, the fact is people still like to be around each other. Given a chance, they still attempt to gather in a relaxed, informal way, in a pleasing, unstructured, unsupervised, unplanned public place.

<hr>

The problem of access to water, or shoreline, has become a major private property issue in those states that once belonged to the Massachusetts Bay Colony. The old colonial laws regarding property along the coast, which still prevail, held that an owner's rights extend down to the low-water mark, and unless you are fishing or fowling, you cannot walk on privately held shores. The original intent was to encourage the owners to build quays for trading purposes, essentially to encourage commercial development along the coast, but in our time, with shrinking open space, and the seemingly deep-rooted human need to spend time at the water's edge, the private property issue has reared its head again and again along the beaches of the Massachusetts Bay Colony states: New Hampshire, Rhode Island, Maine, and Massachusetts.

On the south shore of Martha's Vineyard, which has a twenty-two mile sand beach and offers excellent strolling opportunities, private landholders in some sections have resorted to posting guards on the beach to keep people off in summer. One friend of mine from Switzerland, who was visiting the Vineyard, made the mistake of presuming the beaches of the island were free and early one morning took himself off for a walk. He parked his car at the Katama landing on the south shore and headed west along the wide expanse of beach. The wind was at his back that day, the sun was rising behind him, and the whole world seemed empty and wild. A mile or so into his sojourn, he heard a shout from the dunes. A stubby man was there with two Dobermans on a short leash. He waved at my gentle friend, who, presuming innocence, waved back. The stubby man shouted something else and waved some more. These latter waves were less amicable, but my friend carried on. The next thing he knew, he heard the huff and cough of straining Dobermans at his back, and the stubby man accosted him. There were many profane English words which my friend missed, but the intent was clear. There had been some violation, and he was forced to return whence he came. Later over breakfast we had to explain to him this concept of private property.

The poor man. We told him to walk left at Katama on what was public beach, which the next day, he did—only to be nearly run down by off-the-road vehicles. Such is the dilemma of open spaces in America: it's either feast or famine—a private peace, obtained by the rich at great expense, or a motor-infested, noisy, littered public space.

Elsewhere in Massachusetts these colonial laws have led to suits by those attempting either to close or open public access to shorelines. Along the popular Atlantic Path in Rockport, a private landholder, exasperated by abuse of his property at the hands of wandering sojourners, blockaded the property to prevent passage. The land in question had been open to walkers since the 1920s, but because the judge determined that use of the trail along the shore was irregular and that, by unwritten local tradition, people avoided passing the house when the family was at home (signified in this case, not incidentally, by a raised American flag), the judge upheld the right of the property holder to exclude walkers. The local free-passage community went up in arms. This surely would be a seminal case, they argued, and sure enough, a few years later, another private property holder in the area blocked access to *his* section of the Atlantic Path.

These of course are minor skirmishes compared to the private property wars of the American West, some of which involve shoot-outs, killings, and prolonged sieges by federal officials. The issue of private passage has also reared its head in England, where rights of way have been maintained over the centuries, long before the concept of land as property had even fully evolved. In recent decades, private landholders have been attempting to close down these traditional paths. In reaction, an organization known as the Rambler's Association has formed and periodically holds what are termed mass trespasses, in which demonstrators purposefully and quite publicly walk out over recently closed sections of common land and public ways to

assert their ancient rights.

As usual, in all these cases, it's not a simple question of who is right. In Rockport, the landowner claimed that the public was throwing litter on his land and peeing on his property in full view on Sunday mornings. In England, on the moors, landholders argue that Sunday ramblers are leaving the traditional paths and wandering out over the moorland, crushing rare plants, stepping on moorland bird eggs, and dispersing rare nesting birds. The problem is really not a question of rights of way, but of manners. Farmers argue that your modern-day walker, with no knowledge of country ways, leaves gates open, allowing cattle to stray, and here in America, owners argue that the public is loud and abusive, carries noisy radios, and drives big machines that chew up the earth.

What most people really want is not necessarily to own property, but to be assured of a certain peace, as witnessed by the archdefenders of our landing—we who do not own an inch of it, but rise up to defend it vehemently whenever its peace and quiet are threatened.

One of the most interesting solutions to this problem of protecting desirable private tracts from an abusive public was the action taken by private landholders near Eureka Springs, Arkansas, where a pleasant rural pull-off was a popular parking and gathering site for the locals. The new owners of the property were horrified to find how abused and littered the land was at that site, but rather than post No Trespassing signs and attempt to block the public, the family decided to honor the place and actually encourage visitors. One idea led to another and they ended up hiring the architect Fay Jones to design a chapel for the site, to emphasize the spiritual beauty of the wooded area. Jones proposed a design that would include rather than exclude the natural surroundings, and he set the chapel away from the road at the end of a wooded path. The place now has become a pilgrimage site and has more visitors than ever before, but no one deigns to litter the site, and people come away with an appreciation of the landscape.

❦

The Long Lake landing has a diverse natural habitat, with dry uplands, a marshy shoreline in one area, rocks in another, and lovely hidden coves overhung with old oaks and beeches.

In one section, on a soggy bank just west of the landing, there is a healthy patch of groundnut. This twining vine, which puts out strange chocolate-colored flowers, has tubers strung along its root system, connected by narrow, fibrous strands. On some plants you can find as many as ten tubers, the size of small, new potatoes, and since groundnut tends to grow in profusion, you can dig through an area, collect a fine dinner, and come back the following year and find even more tubers in the loosened soil.

Sarah Doublet herself may have collected these plants at this very site. Groundnuts were a major staple of the Eastern Woodland people, and she and the other women in her band would certainly have known where all the good stands grew. I can imagine her wandering over these shores with her sisters, pulling up the vines to check the stage of the tubers over the course of the summer. During the growing season of 1676, she might have paid especial attention to the patches of wild plants that sustained her people, since the corn crop was certainly diminished in that sad year.

By February of 1676 there were 550 Indians incarcerated on the twelve-acre Deer Island. Curiously, like the Nisei in their internment camps, many of the Christian Indians on Deer Island had remained loyal to the English in spite of the harsh conditions. The men were actually anxious to help in the war against King Philip. Part of this, of course, may have been a stratagem to escape the harsh conditions of the camp.

By that winter, the war had begun to turn against the Indians. King Philip was short of food and needed allies to continue his battle against the English, and so he sought the support of the Mohicans along the Hudson River near present-day Albany. But the old animosities between the Indian tribes were still a reality, and it did not take much persuasion by the governor of New York to convince the Mohawks to attack the Mohicans. The Mohawks were victorious and subsequently pursued King Philip and his allies eastward and, even after the war ended, continued to harass the eastern tribes, as they had for centuries before the arrival of the English.

In spring, it was clear that the English would win, and in May the Massachusetts General Court ordered the release of all the Indians held on Deer Island. Many of the Christians returned to Natick, while others simply dispersed into Boston and its environs. Some ended up as servants in colonial households; some attempted to return to their old way of life. But, except for Natick, where they held onto their ancestral lands longer than they did in the other Christian villages, the loss of their old landholdings and the continued threats of Mohawk raids, plus the lack of a community to carry on the traditional farming, made a return to the older way of life all but impossible. Many simply took to the bottle.

The Nashobah people, or what was left of them, returned to Concord and worked for local families. But a few, including Sarah and Tom, and possibly the man called Crooked Robin, John Thomas Good Man, and perhaps John Speen, came back to the old village site. It must have been a sad return. The old planting grounds would have been covered with weeds. Somewhere in the village, either stored inside one of the houses or buried in the earth, they would have left baskets of dried corn, and this they would have searched out and ground for meal. But the place was emptied. Mice nested in the dry walls, and the mats lining the wigwams and the bones of whatever game they had left behind would have been gnawed clean. The New England spring—the fresh greening hillsides, the flowering of their apple trees, the night croak of frogs, and the dawn chorus of birds—mocked their poverty.

Did Sarah still pray, I wonder, after all the abuse she and her people had suffered at the hands of Christians? Most likely she mingled the two faiths with even more fervor. She hunted birds and braided their feathers into her long black hair. She once again tied the moosehide thongs around her ankles, tied her waist with blue trade cloth decorated with glass beads, threw her blue mantle over her shoulders, circled herself with strings of wampum, and fixed the horned water beast amulet to her neck. Perhaps she mourned, as she would have at the deaths of her various husbands and her son. She ground clays and painted her face black. She sat in front of her emptied wigwam, and took the cold damp ashes from past fires and rubbed them on her cheeks, twisted her hair, and with a stone or steel knife once more cut it short. She began rocking, chanting the old hymns, her hymns, mixed with their hymns; god of our fathers, our Father which art in heaven, *Newutche nag Jehovah Manit*— Heavenly Father god Jesus Christ his son have mercy Jesus Christ—*Lord num Manittoo mun Jesu Christ.* And maybe she sang out with a long traditional howling that echoed off the rock walls behind her and the pond waters below and sailed off into the wider world. And maybe the earth opened again in that sacred place and the very rocks sang in the upper air, and the four winds inside the hill opposite her began once more to growl and snarl and sing back to her, Sarah, the guardian spirit of this spot. And maybe the earth erupted again as it had ten years earlier and left a great hiatus and the humming noise issued forth from the ground and the hard earth rose in a wave like the sea. Through her tears she would see the torrents falling, and a rain of trees and branches, and the solid earth would roll again into a vast swell and settle, and vapors and contagions and afflictions would spew forth from the place that was Nashobah, the smell perhaps of sulfur, of Him that rides by Night and the beings of the invisible world, the fairies and demons and devils so well known to Cotton Mather would come winging overhead in dark swirling flocks, the black man and his horde, swimming in the smoky air, homunculi with clawed feet and doubled legs, red-eyed, with bat ears and pointy little fangs, carrying in their right hands the forked fish spears with which to prod downward into hell and fires eternal this known sinner Kehonosquah, also known as Sarah, Sarah Doublet, or Sarah the Indian.

Maybe she stood at this point and ripped back the blue cloth mantle to bare her breast to them and then she stepped forward, once, twice, and back once, twice turned, stepped forward once, heel and toe, a shuffle, and then the howling began again in earnest, the mix of languages, and who knows how long she danced in that place, if she danced at all, until finally, exhausted, emptied, drained of all her history, all confusion, all physical energy, she fell into the emptied wigwam, wrapped herself in her blankets, among whoever else was there with her, and dead to all feeling, after a winter of nothing but feeling, she slept.

Or did she simply walk out the trail from the Great Road, turn, as Mosely had, the autumn before, at the southern shores of Nagog, and skirt the banks over to the old village site, where she climbed the hill and sat there in silence, not rocking, not howling, not praying, but simply staring, her old spirit dead within her, a numbness, a lack of feeling?

Inasmuch as she came back in planting season, it is possible that Sarah, aided by whatever women were with her, may have cleared the ground of weeds by burning, and then raked the detritus into the hills. They would have missed the herring run that year, and so they had no fish to fertilize the hills, and without the boys to guard the seed corn from the crows and blackbirds, it is unlikely that much of the corn planting took, if indeed it was even planted. Sarah and some of her children, if there were any with her, would have set up a wigwam at the planting ground perhaps to guard the crop and as the young shoots appeared, they would have tended them, as they normally would do, until the shoots grew into stalks and the stalks put out ears. And if indeed there was a harvest, it would have been a sad event with thin pickings, and since there were few people left at Nashobah, there would have been no Green Corn Dance that year, no celebrants, none of the howlings that passed for singing, no drunkenness, no ecstasies.

The apples, by contrast, may have done well on their own, and Sarah would have collected the crop the following autumn, pressed the apples and made her cider, which she would have fermented and perhaps even distilled for later consumption. Now though, there was more rum in the region, and drinking was no longer a ritualistic, ecstatic celebration, but an escape. That next winter, according to court records, there were complaints of drunkenness among Indians on the streets of Boston. One of them was found dead with a bottle of rum propped between his knees. Slatternly, drunk Indian women were charged with vagrancy. People lived on the streets or just beyond the town in rough wigwams. They traded baskets and berries and bearskins, and some of them, no doubt, professed Christianity, and could mumble prayers and still sing their version of Christian hymns. But even in the former Christian town, in Natick, there were complaints by pious English visitors that the heathen religion was creeping back into the Indian services.

The Indians were by this time under the strict control of the English. There are records, for example, of Indians petitioning the General Court to gain legal permission to go to a certain hillside to collect hurtleberries, and there is an interesting petition seeking permission to sell bearskins in Boston.

Where the Indians of Nashobah were actually living after the war is not at all clear, but the presence of the fort that gave Fort Pond its name gives some indication. The stockade–it would not have been a fort in the English sense–was a defense against other Indians, not against the English. After the inducement to raid the eastern Indians, the Mohawks had taken up the challenge with relish, and began, as was their custom before the English arrived, making forays eastward. They raided Christian and so-called "friend" Indians indiscriminately. They carried off two older women who were making cider at the nearby town of Hassanamisset. They captured twenty-two Christian Indians from the Natick community, who were working in the fields in the Christian town of Magunkaguog, and they even killed two Indians at Spy Pond, in Cambridge, only a few miles away from the General Court. The people at Nashobah must have been concerned. But the fact that a fort was constructed in the area, plus the traditional name, Speen's Point, for the jut of land at Fort Pond–not to mention the presence of an old road west of Fort Pond called Newtown Road–suggests that there were more Indian people around Nashobah than the records indicate.

After their return from Deer Island, Sarah's husband Tom went up to the old fish weir on Beaver Brook and built a hut for himself, about three miles north of the village, and lived there alone.

Local histories of the English town that was established at Nashobah in 1714 make mention of Tom, although there is no mention of Sarah. But then

English historians and record keepers tended to overlook the presence (not to mention the importance) of women among the Indians. An English family named Proctor had built a house in the area by the 1680s, just upstream from the spot where Doublet's weir was located, and the family reported that Tom was a tractable old Indian who spent his last days catching fish.

Sometime in these years, the histories do not say when, Tom Doublet died. He does not appear to have had any vested interest in landholdings in the Nashobah area, and after his death, under the English system, any rights to the tract he held would have passed to Sarah. Under the native system, she would have already held them if she was, as Chuck Roth thinks she was, the great-granddaughter of Tahattawan.

If he had a Christian burial, which is questionable, Tom was interred in an unmarked pauper's grave. Sarah or any other relatives may have set up a stone at the site with no lettering, no indication of his individuality. When the Nashobah people buried him, they may have once again reverted to their old religion and sent him off to the spirit world with a few grave goods. They dressed him in a prized doublet awarded to him for his work as a translator during the war, and placed him in the grave with some brass buttons and, perhaps, the fish spear he used to get his eels. And then they covered him up, casting in with the soil a little red earth if they had any around, and perhaps mumbling over him a few prayers in English and also in Algonquian. Then they walked away, and after Sarah died, he was forgotten.

One by one, year by year, the original inhabitants of the Christian Indian village disappear from the records. These deaths are not recorded. Essentially, the Indians, even though they may have held power or position within their own community, if they were recognized at all, were recorded in the historic documents as paupers by the English and then only because towns who had assumed their care were seeking recompense from the General Court for their expenditures. Cows, sheep, and pigs are better documented in the form of community tax accounts.

If there was a leader among the few people who were living around Nashobah at this time, it would have been Sarah. But she could also have inherited Nashobah simply by attrition—she outlived all the others. In the end, she was the last to know where to find the good stands of groundnut. She knew where to find Indian tobacco and the sweet flag and the edible wild greens. She was the last to know the good sites for fishing weirs, the location of patches of goldthread, the last to know the healing barks and roots, the medicinal plants of the five-hundred-acre tract around Nashobah.

Nevertheless, somewhere around the Nashobah tract, at least during certain times of the year when the groundnuts were full or the blueberries on Nagog Hill were ripe, the remnant Christian Indians may have come together for ceremonial purposes. And on certain nights, perhaps in August, when they would have held the Green Corn Dance of old, they sat around the fire, and as lost people will do, reminisced about the time before the time that the *Awaunaguss* came among them. And on certain nights it is possible that the black wall of the woods closed in on them, and the moon cleared the cloud rack, and the old magic began to stir. And it is possible that on some such night, Sarah rose up and dressed herself in English clothes. She drew on a kersey coat and a string of wampum and whatever beads she had found after her return. And sitting by the fire, she mixed the clays and ground them, and once again painted her face—half black, half red—the black for mourning, the red this time for celebration. She pulled bracelets onto her arms, from wrist to elbow, and tied the deerhide strings around her ankles as her mother would have done, and set the old swansdown earrings in her ear lobes and pulled on red English stockings and white shoes, and she powdered her hair white. And all the others, seeing this, pounded the ground with their flat palms, and Sarah rose and stamped hard with her right foot, and then with her left, and then the great circle began. She shuffled forward with her arms cocked and they slapped the ground, and then she turned and shuffled backward, and they hit the ground. And then a rhythmical chanting began, and the others, who were dressed also for the occasion, got up to dance. They wore Holland shirts and silver buttons and brass-buttoned doublets and waistcoats, and English shoes, and all of them were strung with beads and wampum and their faces painted black or red, and their hair done up, if it was long enough after all the mourning they must have done in those years, in the elaborate coifs of old. And they set a cauldron of warm water on the fire and danced around it, quenching their thirst as they circled. The moon crossed the sky, and, this being late August, the migratory nighthawks floated over the treetops, heading southward, and the black wood drew closer with all its mysterious yelps and rustles and barks.

And maybe down at the lake, the long-banished water beast rose to the surface for the last time, and slid his glistening tentacles out along the rocks of the shore, as he had of old, searching for victims, and maybe that night Keitan came up from the forest, and Hobamacho, and Bear, and all the other demons from their pasts, the nightmares that haunted the sleep of their childhoods, in the time before Jesus Christ and God walked among them and the living nightmare of the end of their culture came to them in the form of "these strangers," these *Awaunaguss.*

Or maybe they all danced together that night—Jesus, Keitan, Bear, the Holy Ghost, Hobamacho, Wolf, and Turtle—all emerged to form the great circle that was once a part of their life, the all-inclusive pantheon that would soon be replaced by whiskey and poverty and drugs and the profound loss of spirit that would characterize the American Indians after they lost their common ground.

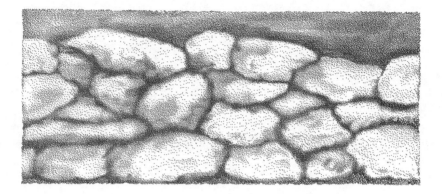

Chapter Eighteen

[Because of] the death of many and remmovall of [others], who during the time of the late wars have been Sojourning among the English for their support, and are not yet returned to their plantation, we are now greatly deminished & impovereshed. [Our] meeting house where wee were wont constantly to meet Sabath days and lecture days to worship God is fallen down and we are not able to build us another.

–From a 1699 petition to the General Court by Christian Indians.
Quoted in Jean M. O'Brien's Dispossession by Degrees, *1997*

Drawn and Quartered

By late February the whole orchard had taken on a gloomy air. There was a heavy snow in late winter that split many trees, and after that, it rained in the hills and began to warm so that there were a few patches of open ground in sheltered areas. I walked the lands of the Sarah Doublet Forest bundled in sweaters and scarves, Wellingtons and corduroys, and imagined an English spring with green lawns and old oaks clawing at the leaden sky.

On one of those winter walks, in a moment of bravery, it struck me that I really should just go ahead and introduce myself to Raging Bull and have it over with. I used to see the old man from time to time that winter and generally avoided him, except for the time that I dared to greet him. But what to lose by knocking on his door and greeting him formally, and explaining my interest in the history of his land? He was not a keeper of killer dogs who would attack on sight, and I presumed, were he to come armed to the door, he would at least have the courtesy to ask first and shoot later. I believed I could make a good case for my presence, since both he and I obviously shared an interest, I might even presume to say a love, for this anomalous plot of land.

This was one of those steely gray winter afternoons in February when the pale winter sun barely gleams through the cloud cover and the white ice of the ponds and the gray-green grass of the orchards cast an eerie pallor over the landscape. There were two ash trees on the south side of the house, and three crows were hunched on one of the lower branches, their beaks pointed eastward toward the pond and the Isle of the Dead. I pulled off the road and proceeded down Morrison's brick path, past the clipped yews, to the formal front door. There was no bell, and the knocker was sealed behind a locked storm door. Obviously this was not the main entrance, so I went around the side of the building, feeling very much like an intruder at this point and beginning to lose my nerve.

But from time to time in a peasant's life he must perforce go through such ordeals and approach the lord with petitions, appeals, and applications, so I walked along the side of the house, to the kitchen door, passing en route a window that revealed a hallway with a washing machine and the kitchen beyond. Then steeling myself against adversity, I knocked.

No answer.

I tried again. No one appeared. And then after the third try, somewhat relieved actually, I gave up. On the way to the car I saw the manager for the property and asked him if Mr. Morrison was around.

"No," he said. The manager rarely gives more than monosyllabic answers. But I persisted, as one must with this man. Will he be back? I wanted to know. No. Will he be around tomorrow? No. I wanted to talk with him. Can't. Why not, is he out of town? No.

"So what's happened?" I asked.

"Sick again. Hospital."

Three weeks later I saw his obituary in the local paper.

The Lord of the Nagog Hill Orchards was not the only landholder to die during that time at Nashobah. In the autumn, Dorothy Flagg, Farmer Flagg's mother, had been taken to the local rest home, as Mr. Couper calls it, and there, in May, she died.

"Mind was gone, though," her contemporary and good friend Mr. Couper said. "'Bout time the body caught up."

With Morrison dead, and Dorothy Flagg gone, and Chicken John in his grave, Mr. Couper was now the last large landholder of his generation in the Nashobah area. His cousin Junior Kimball, the eighty-year-old marathon swimmer, lived on the other side of town and preferred to keep a low profile in land matters. I was worrying a lot about Mr. Couper. Sometimes I would stop by to see him and be told that he was napping or that he was not feeling well that day. Winter seemed to hold him indoors that year, and when spring came he did not reappear as regularly as he used to. One cold day I saw him walking to town dressed *à sa mode* in layers of multicolored rags. He had draped some sort of hood over his head that flowed down around his shoulders, and he was walking tiredly, with the aid of a long wooden staff. He looked for all the world like a medieval pilgrim on the road to Santiago de Compostella.

Unlike his contemporaries, however, when the weather warmed, Mr. Couper emerged from his dormancy. I thought I saw him one spring day at the farm stand, and I pulled over and walked back to chat. He was nowhere to be found. Then I heard some grunts and expletives coming from underground and saw a ladder emerging from a hatchway behind the stand. Mr. Couper was down the hole, his home-dug well, retrieving a wrench he had dropped while setting up a pump.

I steadied the ladder for him while this eighty-four-year-old man scrambled down and rummaged in the slime at the bottom of the well. He came up covered in mud, dressed in some species of garish plaid pants that he had cut off just below the knee, a blue wool sweater, and fashionable horn-rimmed glasses, such as a literary man might wear. He invited me in for a sandwich and told me about an upcoming celebration that had been organized by the Children's Hospital in Boston.

He had recently made a donation of $50,000 to the hospital to honor a certain Dr. Beth Vincent.

I asked him about Dr. Vincent, never having heard of the woman.

"Well, she operated on me eighty-one years ago," he said.

(One never knows what generous thoughts lurk in the long memories of New England farmers. I used to know two old bachelor brothers who donated their five-hundred-acre farm to the Massachusetts Audubon Society because one summer afternoon back in the 1920s, when the two were teenagers, a visiting naturalist told them the property would make a good bird sanctuary.)

Even before the fate of the Frost Whitcomb tract was decided, a notice appeared in the *Wall Street Journal* advertising a farm estate with existing orchards, in an undisclosed location, to be sold, as the ad made clear, as an agricultural operation only. It was Morrison's land.

The news of Black Jack's death had inspired much gossip and comment in the town, as well as an Op-Ed piece in the local paper by a columnist who, not unlike myself, had never actually met Morrison but had been an observer of him from afar for a number of years. He was, if nothing else, one of those individuals who, sometimes unintentionally, ends up shaping the character of a place by his very presence.

The columnist wrote that she had known Morrison only as a farmer, an orchardist, and was surprised to read in the obituary that he had been a member of the Safari Club, and the Boston Ancient and Honorable Artillery Company, a group of established males who attire themselves in militia uniforms, fire cannons in the Boston Common once a year, and accompany the governor and other luminaries at celebratory events around the city. She did not know he had built the low-income housing at Columbia Point in Boston, did not know he had constructed buildings at Boston College, that he had served in the navy in World War II. She only knew the curmudgeonly old farmer who commanded the hilltop at Nagog Hill with a meticulous attention to order and through his efforts created the billowing white clouds of apple blossoms that graced the hills each spring, scented the air with perfume, and sent their petals drifting across the roads. She knew only of the old man who brought busloads of inner-city children to his orchards each autumn to let them gain the experience of understanding, firsthand, where apples come from, and that this same man went his own way when it came to town regulations and commandeered the local crossroads with his slow-moving golf cart.

The main gossip in the town had to do with the fate of the property. Whatever was to become of this gracious land, now that its lord and protector was dead? In nearby Acton, the Morrison estate decided to sell the tract by the Ice House Pond, which Sacha and June and the New View cohousing people had looked at and rejected because the old man refused to negotiate the price. His estate was more amenable, it seemed, and in the spring, after the affairs were in order, the town of Acton, without much opposition, voted to buy the tract to add to its own conservation lands. The land connected nicely with a series of properties the town owned, and had footage on Ice House Pond, which the town had cleaned up the year before. Over the hill, on the east side of Ice House Pond, and east of Nagog Pond, there was another town conservation tract, and here, at the end of a row of anomalous stone walls, which Mara and the antiquities people believe to be part of the ancient Indian sacred lands that focus on Nashobah Hill, there was the sacred (so-called) cave or rock shelter known locally as the potato cave. Here, it was argued, native shamans would meditate on the solstice events, to assure the continuation of human existence on earth through their ritual celebrations.

That summer the bulldozers moved in on the Beaver Brook property. The designers left two handsome old oak trees standing beside the road at the entrance to the development, and then began blasting and stripping their way through the former orchards, hayfields, and pine woods down the slopes toward the banks of the brook. For weeks the scrape and roar of machines filled the air. Raw earth was piled here and there around the tract, new mini deserts were created, which baked in the hot summer sun. Hideous tangles of uprooted trees and brush were pushed into browning piles. The leopard frogs, the wood turtles, the garter snakes and the red-bellied snakes, the bluebirds, the song sparrows, the towhees, the prairie warblers, the yellowthroats, the buck moths, and the snowy tree crickets, which once held sway in the one-hundred-acre zone, were all dispersed, either that or killed outright.

My house is not far from the tract, and for weeks after the bulldozers moved in I could hear the sad descending whinny of a screech owl wailing from the surrounding woods. No doubt it too had been evicted from its territory. Mara, who heard it calling one night, claims it was crying for a lost land.

But except for the Beaver Brook development, it could be argued that things were going well for common land around Nashobah, although the fate of the very heart of the old Nashobah tract, the Morrison property, essentially the last undeveloped Christian Indian village site in New England, was still up in the air. And since there were no immediate offers, either from private farmers or from any of the national or local land trusts, it appeared that its future would be unknown for a long time to come.

One April day after Mr. Morrison died, I went down to South Acton to see if Mrs. Smith knew that her old house was about to be put up for sale. By this time, with individuals of her generation passing one after the other in swift progression, I was becoming sensitive to symbols, and old Mrs. Smith's house had been looking ominously deserted. I pulled over and walked up the path to the front door.

Mrs. Smith was deaf, and it's possible that she didn't hear my knock, but I looked in the side window and things looked unlived in. I didn't dare ask around to find out what happened. Instead, I drove up the road another half mile and went to visit my friend June at the New View, which was set-tling into the land very nicely, it seemed to me.

This was as fine a day as one gets in April in New England—sunshine, blue skies, and the daffodils and forsythia glaring against the fresh wet green of the lawns. It was also a holiday and by rights everyone should have been outdoors enjoying the weather. But on the way to New View, there was a curious emptiness to the landscape. I had driven up from Lincoln, through Concord and Acton that day, taking back roads all the way, and I was struck by how few living souls I saw out in the well-landscaped suburban yards. I counted maybe five people all told, and they were mostly working alone in their gardens. It was as if some mystic plague had passed over this land and taken everyone in their sleep.

There was a different scene at New View. I pulled in and parked my car at the common parking lot and walked down the hill toward Sacha and June's small house. There are some thirty children living in the community and they were everywhere that day, some skating, some bicycling, some running here and there in small herds. There was a group of girls standing under the trees gossiping or plotting; they looked as if they had some definitive plans afoot and kept glancing up the hill at one of the houses. A boy who appeared to be Native American, probably adopted by one of the residents, scooted past on a bicycle with training wheels, smiling proudly. There were three women sitting on the steps of one of the houses, chatting, one of them lying back with her face to the sun, and all through the common areas there were people outdoors.

Just north of the common five or six men and women were working in the area set aside for the community gardens. Some were spading the ground, some were laying out onion sets and seeding the rows. June herself was working with a neighbor and two little girls to put a new vegetable and flower garden in what, in a traditional feudal village, would have been the homelot, the small privately cultivated ground near one's dwelling. It all seemed so healthy, the small town reborn, everything that Lewis Mumford and Henry Stein had in mind when they began designing the new towns for America and used the image of the common at Shirley and the old New England land use pattern as a model for the future. But this was an anomalous day at New View, a holiday. No one actually works there, as they would in a traditional small town.

This happened to be Patriots' Day, a celebration of that day back in 1775 when the colonials, upset that their inalienable rights were being abrogated by the sovereign lord back in England, rose up in arms and gathered at the bridge in nearby Concord to fire the shot heard 'round the world. Tomorrow, all the New View residents, like every other suburban American, would get back in their cars and drive off to work and spend the requisite number of hours in their artificially lighted offices and then return at night, when it was too late to work in the garden or gather in the common. New View worked best for the children, it seemed to me. They actually lived here, side by side with their friends and allies and enemies. They played outdoors, relatively unsupervised, little free-ranging packs, sent off to explore the world on their own. If nothing else, New View would be an excellent place to grow up.

Dennis and company came back to the orchards that spring and began work as usual, even though there was no old man to boss them around. In his place, the manager of the property, who drove around in a late-model gray Mercedes instead of a golf cart or old truck, was running the orchard for the estate. One of the first things they had to do was close down the pick-your-own operation and sell wholesale only. Morrison's second wife, a thin, agreeable woman, maintained, as she always had, as low a profile as possible and allowed the attorneys for the estate to handle the legal matters, and the manager to handle the orchard.

Without its raging bull, the hillside orchard lost something. It was now less threatening, to be sure, but it was also less interesting. Wandering around the place somehow had less of a thrill. I began stopping often at the main barn at lunchtime to look up Dennis and learn what I could of the fate of the property. But of course Dennis knew nothing, inasmuch as he was merely a seasonal worker, and the manager was, as I pointed out, a man of few words. The longest sentence I think I ever heard him say was "I'm a busy man."

This last was indeed true. He would work from sunup to sundown even in summer, when the days were more than twelve hours long.

Late one summer evening I went out to the point of land where the summer cottage used to be located and pushed through the sweet pepper bush to a flat rock that jutted out into the pond. Ahead of me were the towering pines of the Isle of the Dead, absorbing all light even on this green evening. To my right was the busy eastern bank where the Great Road lay, and to my left I could see the long sweep of field that ran up to the old Morrison house, with the apple and peach trees on either side. As I watched, the light began to change. The greens deepened, the blue overhead took on a richer hue, and the yellows and pinks of the western skies behind began to filter behind the trees of Nagog Hill. Somehow, as the sun shifted, a flare of yellow spilled over the white of the house and illuminated it for a minute or two; I looked away for a second, and when I looked back the whole landscape was green once more.

Out on the pond the wind dropped. The gulls began to settle, and the deep reflection of the trees and the obscured horizon detached the Isle of the Dead from this earth so that it seemed to float above the waters with no connection to lake or sky. I could understand why so many people who know this plot of land believe there is something sacred about it. It struck me suddenly, and perhaps not irrationally, that Sarah was interred out there on that island.

One wonders why landscapes of this sort cannot be preserved for totally irrational, uneconomic reasons—saved if for no other reason than because they are there.

A few days earlier I had set out on a walk with some friends through a recently protected wooded area when the father of one of our party materialized out of the forest. He was dressed in old khakis and work boots and was carrying a well-aged cherry walking stick, which accompanied him everywhere. This old man, in spite of his advanced years, was still active and was forever developing new theories and taking on new causes, the latest of which had been a campaign to save this particular stretch of forest. He was not one for small talk. Spotting a stranger in our midst, much to his daughter's chagrin, and without prompting, he launched into one of his speeches. He fixed the young man with his fierce blue eyes.

"We can't own land," he pronounced. "Doesn't matter what the law tells you. In the end, we are just sojourners. We are merely the stewards."

⋙~⋘

I met a new Jamaican at the place that summer, a tall, heavyset man with a pleasant face. I saw him nodding at me from the background one day as I spoke to Dennis, as if he wanted to join in the conversation, and so later I sought him out. He was a talkative, easygoing fellow, but as he said, he had never known this old man and had no idea what would become of the land now that the lord, "dat ol' mon," as they called him now, was dead.

In feudal times, all would have been spelled out quite clearly by the cultural traditions. The peasants, be they serfs, villeins, peons, or thralls, belonged to the land. If a new lord came and assumed control of the estate, he was in essence buying the peasants as well as the land; they themselves had no say in the matter. They could not get up and leave, they could not own even the homelots that they worked privately, and there was no way they could roll their earnings (there were no such things as earnings in any case) in their socks and carry them home to the native place. They were chattel, not too much different than the cows and sheep of the estate.

But this question of belonging to the land, of a connection with a piece of earth, also had its advantages. The peasants were an integral part of the economic system in Europe and, under the allodial system, had unwritten rights to be in that place, no matter who was king or lord. The system was (still is actually, to some extent) in operation in the Spanish colonies. The peons remained on the land even though they didn't own it. But this was not true under the unique system of private property that was fast evolving in the English colonies, and so Sarah, who essentially could claim legal ownership of her five hundred acres, could also sell—and thereby lose—that same piece of earth forever. Her heirs, whoever they were, were about to be disenfranchised.

<hr/>

By 1680 Sarah Doublet had been living around the former village site for some four years. She was now in her forties, had been married at least twice, had lost at least one child, the twelve-year-old killed by the English at Wamesit, and may have had one or two, perhaps three, more children. In London at this same time, the city was emerging from the ashes of the Great Fire and reshaping itself from the drawing boards of Christopher Wren into the city we now know. In the mornings, the people, those of quality at any rate, met in Paul's Walk to pass the time, promenading to and fro in their finery and exchanging the gossip that became, in the words of one of them, the great mint of famous lies. From thence they moved on to the recreations: the cockfights, the theater, or to play at paille maille and bowls in streets set aside for such games and divertissements.

Henry Purcell was at the Abbey of Westminster on the console of the great organ in those years, and the gentlemen of this city went to the barges on the River Thames and rowed to Chelsea or up to Deptford to have a drink on board Drake's own ship, the *Golden Hind.* The chronicler of these comings and goings might have been there too, for all we know, Mr. Samuel Pepys of the Royal Navy, with his gossipy diaries. The ladies went to the taverns in masks, and there was much gambling and losing and gaining, and some went to musical water parties and some went to the Sadler's Wells entertainments at Islington, and some, some few, went about in curtained coaches, courting, and perhaps dicing with the men at the taverns, and although much abused by the Puritans, everyone, men and women alike, went to the plays of Beaumont and Fletcher and Dryden. Henry Purcell set Dryden's "Indian Queen" to music for them and put on a great tragic masque with spirits singing in the air and the god of dreams ascending through the trap and battles and human sacrifices on the very stage.

The Puritan victory in the English Civil Wars and the rule of Cromwell and company had not entirely rooted out the pleasures of the court. The fops still gathered in Hyde Park and Mulberry Garden, and there were may shameful gallants of both sexes in powdered hair, and there were "*spotted women*" all painted and coiffed. As Mr. Pepys, somewhat hypocritically (he was among them) points out, the times were in ill conditions, and the vices of drinking and swearing and loose amours were so well established that poor Mr. Pepys (so he says) knew not what would make an end of it. All this dancing and drinking and quaffing, and even Cromwell could not (preferred not in fact) to quash the music. He may have looked down upon the masque, but he allowed the beginning of English opera and so Mr. Purcell himself stepped down from the organ console and composed another famous masque, "Dido and Aeneas." And then, after Charles, whom they slew, England got another Charles, and the Restoration was in place.

But you would not have known much of this from the cold shores of New England. Theater was illegal; wearing your hair long was illegal; adultery was illegal. Taking the name of the Lord in vain was illegal. Wearing a buckle upon your black Monmouth hat was illegal (unless you were a freeman and a member of the church, which was forbidden unless you could pass a series of interrogations which would permit you to enter into the sacrosanct body of the congregation). The Massachusetts Bay Colony was a theocracy. In spite of the brave efforts of Roger Williams, who sought to level things out, there was no difference between church and state, and the state made no differentiation between that which was a matter of state and that which was a matter of social life. Cursing was not a social faux pas, it was a civil crime, and you could be sent to stand in the stocks for it. Dancing, under certain conditions, was a crime, so was acting. The great theater of the day, the social event that drew the crowds, the one public entertainment that was permitted, encouraged even, and apparently enjoyed, was the witnessing of criminal punishments.

By the 1670s public punishments such as a whipping or a mutilation would draw a local crowd. But the main event was an execution. These were announced well in advance, and on the big day the crowds would gather, some individuals having traveled for days to attend. As the moment arrived, the drums sounded and the participants marched in, in procession, led by the black-robed ministers. Long sermons were delivered on the nature of sin before the actual execution. The sermons were graphic accounts of the crimes, and the criminals were encouraged to play their own parts. When they hung the pirate James Morgan in Boston in 1686, crowds began to arrive in the city a week before the day. Increase Mather himself preached the pre-execution sermon to some five thousand people, some having come all the way from the Connecticut River Valley to witness the scene. Vendors sold broadsides recounting the crimes of Morgan, and the grand theater took place at high tide to great gasps and cheers from the assembled when the trap was sprung and the heinous pirate dangled, kicking, from the hangman's rope.

Such was the state of New England in 1680—more or less.

Out at Nashobah there was a sad wind blowing. Indians would come and go from the former village site. John Speen, who had family at Natick, was around enough to get his name mentioned in the local histories. The children of John Thomas Good Man were around; Crooked Robin was dead, though; at least there is no mention of him. But there were still a number of Nashobah people in the general area. Daniel Mandell, who wrote a book about the dispersal of the Christian Indians after the war, says that there were as many as fifty Indians living in greater Concord, most of them working as servants or trading with the ever-richer English families who were farming and trading around the Grassy Ground River, which the English had by this time renamed the Concord River.

These Nashobah people were, as the Indians on the western frontiers would be two hundred years later, something of a thorn in the side of the good Christians of Concord and the other Puritan villages. Indians in the more settled areas were no longer a threat. But the clashes of the two different cultures till grated on the English sensibilities. Indians still had no clear sense of boundary, no demarcation of property, it seems. They were notorious trespassers and cattle thieves. They had the interesting custom of simply walking into a house without knocking and taking a seat by the fire, without a word, it is said. There they would sit, silently waiting for a gift of tobacco or food, and there they would remain. After some time, they might, in their broken English, explain that they had some furs to trade, or some fresh game, or that they would be willing to work for a few days.

This business of regular work was still not something the Indians could come to terms with, nor the Puritans with the Indian ways. Work for the Indians had always been a sporadic affair, and they were still not accustomed to the steady, day by day grind that was so much a part of the Puritan makeup. And so the Indians were labeled lazy. Also dirty, given to drunkenness. By contrast, what the Indians thought of these white people, what they said about them in their private moments, is unrecorded. But certainly, given their traditions, it must have seemed odd to enter the equivalent of a wigwam and not be offered the traditional bowl of food, or a gift, as was standard practice in their culture.

Slowly, year by year, the Indian lands of the greater Nashobah Plantation were sold, stolen, or otherwise transferred to the white settlers. The nearest English settlement, Concord, had begun the process of formalizing its land claims as early as 1653, when the townsmen were busy with the second division of lands of the original grant. That year, the town voted to have the town divided into quarters and have each deed recorded in the county registry and the town books. By the eighteenth century most of the lands of the original Concord purchase were filled in, and as the families grew and split off, the inhabitants began pushing ever more outward. By 1679, there were approximately 480 people living in Concord. By 1706, there were 920, and that number grew to 1,500 by 1725. The outlivers were still required by Puritan law to attend Sabbath, and some families had to walk or ride as many as six miles each way to get to church on Sunday.

In 1662, while the Indians were still living on the plantation, Lieutenant Joseph Wheeler of Concord petitioned the General Court to grant him two hundred acres of land in the Nashobah tract, but his appeal was turned down. In 1686, the first real survey of the plantation was undertaken by Samuel Danforth partly to locate a tract of the Indian lands for the two men from Concord, Peter Bulkley and Thomas Henchman, who bought the eastern half of the plantation for the sum of seventy pounds. Kehonosquaw, alias Sarah, the heiress of John Tahattawan, sachem of Nashobah, is named as one of the Indian grantors of the sale; so is John Thomas, or Naanishcow, the teacher of the Nashobah people, and Naanasquaw his wife. Tahattawan's sister is also named in the deed. Harwood's history of the town of Littleton says that Danforth probably laid out his plan more in the interest of his clients than of accuracy, and that he probably didn't even run the northern line of the plantation, but merely assumed that it was four miles long.

Two families from Groton, Peleg Lawrence and Robert Robbins, managed to cut out a half-mile-wide swath extending two miles, along the northeastern section of the plantation in 1687, just ten years after the war ended. But this section of the tract was much argued over in the courts when the town of Groton tried to lay claim to more of the Nashobah lands. The titles were questioned, and the land was eventually turned over to the town. Lawrence and Robbins were allowed to retain ten acres each of meadowlands in the area.

By 1694 the purchases of the English had cut the plantation in half. Other sales soon followed. Waban of Natick somehow transferred, for the sum of fifteen pounds, a "certain Tract of Land upland, Swamp, Meadow Land, Containing one Quarter part of an Indian Plantation known by ye name of Nashoby . . ." to Walter Powers of Concord, even though these lands were "owned" or at least associated with the two sons of the Indian named John Thomas. By 1714 there were enough English living in the area to incorporate a town, which the English named, quite appropriately, Nashobah. Two years later, they changed the name to Lyttleton in honor of Sir Thomas Lyttleton, the father of Lord George Lyttleton, who later became a member of the British Parliament. The spelling subsequently evolved to the current name Littleton.

But in spite of all these acquisitions and changes, down on the southeastern corner of the former plantation, in the rampant, earthquake-broken hills between Nagog and Fort Pond, where the rock walls rise like crusader castles and the swamps are loud with frogs, old Sarah Doublet still managed to hold on to her last five hundred acres. Where she lived in these years, how she lived, is lost, of course. They were all lost now to history, but I imagine her surviving by her craft. She wove baskets; she picked berries to sell in the streets. She may have traded furs caught by her husband Tom, otter and muskrat and beaver, if there were any left in the area.

Sometime around the turn of the eighteenth century, this Sarah Doublet, this unremarked, obscure, Pawtucket Indian came to the end of her childbearing years. Even a generation earlier, the advent of menopause would have meant that she would have assumed more power among her people. She would have been able to smoke tobacco with the men. She could even have worked in the tobacco garden, something that was forbidden to childbearing women. She would have had more of a voice in the council meetings, more respect as an elder. But as it was, in 1700 among the English, she became even less of a figure, an old crone, more to be pitied than respected. Toward the end, as her compatriots and the first generation of English settlers died off, she became even more obscure. And finally, when there were none of her people left, and Pawtuckets were rare in the region where they had once held sway, Sarah, relict widow of Tom Doublet, rather than Kehonosquaw out of the line of Tahattawan, the *saunk* among the Nashobahs, became simply Sarah, or Sarah the Indian, or Sarah Indian.

They were all dead; there were no contenders to the former village of Nashobah. She alone held the title to the certain tract of land between two ponds.

Epilogue

And it contains one moyety or halfe part of said Nashobah plantation . . . &
this Line is exactly two miles in Length & runs East three degrees Northerly,
or West three degrees southerly, & the South end runs parallell with this
Line: On the Westerly side it is bounded by the remainder of said Nashobah
plantation; & that West Line runs (from two little maples marked with H
for the Northwest corner) it runs South seven degrees & thirty minutes east,
four miles & one quarter; the most Southerly corner is bounded by a little
red oak marked H, the northeast corner is a stake standing about four or five
pole southward of a very great Rock that Lyeth in the line between said
Nashobah & Chelmsford plantation. . . .

> *–Deed for the easterly half of Nashobah to Hon. Peter Bulkley of
> Concord and Thomas Henchman, 1686*

In 1638 a certain John Jones settled in the Concord area. Following the
new English system of dividing lands among the children, this family,
which remained in the Concord area for generations, began to outgrow its
original holdings and had to seek lands farther west, beyond the bounds of
the town. One member of this group, Ephraim Jones, and his cousin, Samuel,
and a man named Jonathan Knight purchased a large tract of two thousand
acres west of the town. Ephraim Jones built a house in the area and, in 1702,
set up a fulling mill on the so-called Great Falls on a stream known then as the

Great Brook. The Jones mill became a successful operation, and so the family built a gristmill at the falls and later a sawmill. The little congregation of buildings became a center, a sort of commercial area and gathering place, and in 1732, on the northwest side of the brook, the family built another house, which served as a tavern and hostelry for the region. Here, daily, the good people of this little outlying community would gather to drink flip and cider, and toddy, and catch up on the news of the day. The tavern was a two-story rectangular structure with a central entry. On the left side of the entry there was a low room with a bar and heavy supporting beams, where the barkeep marked the fare. Just off this was a small room with another fireplace and a door that led down to the cellar, a well-constructed sort of a catacomb with a ground-level entry that opened onto the millpond created by the dam that the Joneses had now constructed to power their mills. To the right of the entry was a more formal living room, presumably private, and upstairs there were two large garret rooms. Here, passing travelers would have bedded down with other passing strangers, men and women alike, in common sleeping quarters. The place was typical of the taverns and hostelries that could be found in the smallest of towns throughout the English-settled region of the Northeast.

This Jones Tavern, where Sarah Doublet last lived, is still standing, as is the original house of Ephraim Jones and some of the mill buildings, which the family set up in the late seventeenth and early eighteenth centuries. The site is just down the road from the New View co-housing experiment and not too far, a few miles, from the village of the Nashobah Plantation. The tavern has a sad modern history. In the early 1960s, the Iron Work Farm, a group associated with the Acton Historical Society, bought the building, had it placed on the National Register of Historic Places, and began to restore it. Then in 1967 a fire swept through the building and burned over one section of the old tavern. The group is still working to complete the restoration. But unlike the original Jones house on the other side of the brook, which has been restored and is maintained and open to the public, the work on the tavern is languishing for lack of funds.

<hr />

One day in the spring after Morrison died, a chill May day when the old lilacs and crab apples outside the building were in bloom, I arranged to go inside the tavern to look around. The man who let me in had to be off on another errand, and so I spent some time there alone, wandering through the rooms, wondering where Sarah spent her last hours. It's a sad place in our time. There's an auto body shop next door, old railroad tracks run past the front, and there is a group of run-down buildings nearby, some with small apartments nestled into them. Here and there at the site, fine old nine-teenth-century structures gracefully descend, waiting to be rediscovered and refurbished, or obliterated. This is not a remarkable landscape, but from my point of view, the area is far more interesting than the new develop-ments around Nashobah, where neat little mini manor houses have formed, with clean lawns and the kind of manicured gardens with pine bark mulch that Linda Cantillon so despises. The photographer Robert Frank, the poet Allen Ginsburg, and Jack Kerouac (who was born nearby in the same water-shed, by the way) would have appreciated the Jones Tavern area. It has that run-down fifties-America squalor that so attracted the Beat poets and their fellow travelers.

There are now two main entrances to the old tavern, one having been added in the nineteenth century. The original leads directly onto a narrow stairwell that twists up to double garret-like rooms on either side of the land-ing. I cautiously mounted the rickety stairs and began poking through the debris and the historical artifacts stored in the garret. There were a few crates with black lettering, describing the contents. There was an old bro-ken rocking chair, a fine antique bureau set neatly against one wall, some broken dishes, and a pile of straw and sticks that looked like an old rat's nest under one of the eaves. The upper window on the west side gave on to the railroad tracks and just beyond, the old millpond, where, I imagine, ducks used to swim and cows would stand waist deep on hot summer days. Now it was a tangle of choking vegetation.

Beyond the pond, the woods began. I could see the little glitter of the millstream running off to the west through the trees. By now it was late after-noon and the shadows were long. I stood at the window for a long time, staring out at the lowering sun and the bursts of wind tossing the trees. Quite naturally, since this is the building in which she died, I started think-ing about Sarah.

In her last years Sarah lived out her time somewhere around Nashobah, smoking her pipe in the sun of fine mornings, and lying at night in a nested pallet of furs and blankets given to her or traded to her by the English. In those final years someone must have been supplying her with food. She was feeble at the end, and blind, and although even in her blindness she probably could have maneuvered her way over the land, and may even have been able to collect a few berries and tubers by feel and smell, it is unlikely that she was surviving off the land.

The story of her end holds that the two Jones cousins who had built the mills down at the Great Falls wanted more land to the northeast to manage the waters that fed the stream that powered their mills. They heard a rumor that there was this old Indian woman living up in the hills between Nagog and Fort Ponds, and that she was blind and not good for much, but that it was she who held legal title to the land the Joneses were interested in. And so the cousins perhaps went up to see her, and found the old woman living alone.

After thirty or forty years around English speakers, Sarah probably knew some local English by that time, and these two cousins, having located the old woman, suggested to her that they would, in exchange for title to her land, take care of her and move her down to a warm room in their tavern at the former Great Falls. This was an old trick. Land-hungry English often encouraged debts, which could later be paid off only with land titles, by offering whiskey or rum or goods to the Indians on credit. The Joneses may have offered Sarah a daily ration of toddy, or rum. They may have had to lure her with other promises of amenities. Or by contrast, they may have been genuinely benevolent souls who wanted to take care of this last survivor of the stormy years which their parents had endured.

However they did it, subsequent events suggest that, at some point, she must have agreed to place herself under their care. And so the Jones cousins came back to Nashobah one day, perhaps with a horse, packed her few belongings for her, and helped the old blind woman up onto the horse, steadied her, and then one of them mounted up behind her, and rode double, it being unlikely, even if she were sighted, that this woman had much experience with horses.

You could presume the threat of winter had something to do with her final agreement to leave her land, and so this may have been autumn and the leaves of the surrounding maples and oaks, as they do today, were turning and drifting down in the little gusts that rattle through the hills of Nashobah. And you could presume that this little troop of three or four souls who came up to fetch the old woman proceeded gingerly through the winding, narrow track that led from Nashobah south southwest to the site of the mills. Even in her sightless condition, Sarah could no doubt still smell the moist earth and leaves of her mother country, and hear, if she still could hear, the little popping calls of chipmunks, which even now sound out the day long in the woods around Nashobah each autumn. And also the steady clop of horses. And the small sad songs of the white-throated sparrows who moved through her territory each year in autumn.

Did she smile, I wonder, not having experience with horses? Did she feel perhaps a certain honor at being escorted in this grand manner, by these seemingly caring Englishmen, whose treatment of her must have contrasted so very sharply with the treatment of her people when last she rode or walked out from Nashobah with an English escort at the hands of Mosely? Was she frightened? Like many older people had she become so attached to her place that even though to stay meant death, she would have stayed? To leave, in her mind, meant instant death. We know only that this small band of people, at some point in the early eighteenth-century walked, rode, or otherwise escorted a poor blind feeble Indian woman named Sarah down from the hills of Nashobah to a place in a newly built tavern by a millpond.

Once there, I imagine, someone escorted her, guiding her at the elbow as one does with unsighted people, to a little back room and there placed her in a bed from which, so the story goes, she would not again get up. I don't know how many years she spent in that room at the tavern. How long she lived there is not recorded. What is recorded, in an oblique sort of way, are the things that transpired there on September 24, 1736.

On that day, the Jones cousins, accompanied perhaps by a scribe and also a solicitor, and perhaps a few others, came into the room where the old Indian squaw was lying. Some of the people there probably didn't even know her name. Certainly none, not even the Joneses, knew that she was probably Kehonosquaw, great-granddaughter of Tahattawan, the man whose line had held land in the valley of the Grassy Ground River for as many as four to five thousand years, and that she, Sarah Doublet, relict widow of Tom Doublet, was the last of his line. I imagine they approached, calling her as they entered her chamber, wherever it was, and she, lying in her bed, called back and perhaps sat up. She must have been approaching the end at this point, and given her age, she was probably painfully frail. She may have been eighty or eighty-five years old. Like some of her legendary compatriots, John Thomas Good Man or Passaconway, she may even have been one hundred.

And so they came forward and sat by her bed and, in whatever language, perhaps tried to explain, and perhaps read her the deed, and then asked in

English, Do you agree? And then they proffered a quill and an inkwell and then, perhaps guiding her hand to the proper place, said if you agree, then make your mark here, and then she, having nothing left to lose, and being at the end of time, reached out her shaky hand and set there her mark.

<hr />

Sometime after that day—the date is, of course, unspecified—we may suppose that the good men of the Jones tavern were drinking at the bar when they heard from the back room a strange sound halfway between a howl and a whine, a creaking, high female voice.

She, Sarah Doublet, Kehonosquaw, may have heard her totem animal speak. She may have sensed the thing that had come for her in the half light of the garret window, she may have heard him sing to her and tried to sing back. And he came forward a few steps, and she heard his yellow scratchy claws on the wooden floor and smelled his meaty breath, a berry smell, an earth smell, the smell of sparks and fire and woods. Maybe she struggled. Maybe he leaned over her and spoke to her in her dead language, she, the last native speaker of the region, and maybe hearing the old words, the words of her people, she settled back on the bolster, and let out her breath, and her lips pulled back in the semblance of a smile.

And then she rose, and went out with him to wherever it is that you go when you die.

At least I hope he came for her. I hope she did not die alone there in that garret room without any solace from her past.

<hr />

And we who are, by reason of English law, the inheritors of this corner of the Western world do not know the full story of those who were the people of this place. We know only that she, Sarah Doublet, relict widow of said Thomas Doublet, did, in consideration of the sum of five hundred pounds in bills of credit paid to her by Elnathan Jones, Gentleman, and Ephraim Jones, did fully and absolutely, give, grant, bargain, sell alien, convey, and confirm unto them and their heirs and assigns forever in equal shares, a certain tract of land.

We know only that if we go to brightly lit local town offices, as I sometimes do, we will be directed to a computer, and instructed to enter the name and street address of a certain tract of land in which we have an interest, and that there, on a clean, cold screen, we will see written certain facts and figures, the book, page, and number, inscribed in the state archives, a

record, a map, an accounting of structures, of square footage of those structures, of acreage within said boundaries, of taxes owed, and taxes paid, and
then we will be instructed to unfold a large blueprint map and there, in
straight lines, and inscribed with numbers and angles, we will see the place
that was Nashobah, where this Sarah and her people once lived.

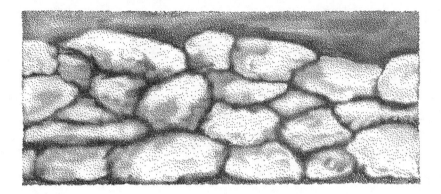

Beginning at a heap of stones in Acton line about twenty rods South of Nagog pond;

from thence running west about Seven rods South Several old marks on Acton line about two hundred and eighty-five rods,

then turning a right angle running north west about Seven degrees and half west by a heap or Row of Stones

where a pine Tree marked,

is fallen down. . . .

—Land grant of Sarah Doublet of Nashobah, September 24, 1736

About the Author

J ohn Hanson Mitchell is author of *Ceremonial Time* and *Walking Towards Walden,* and editor of *Sanctuary* magazine. Most of his published works deal with the idea of place, but before settling on the square mile that is the subject of his previous books, he was an inveterate trespasser.